HYDROLOGICAL PROBLEMS AND ENVIRONMENTAL MANAGEMENT IN HIGHLANDS AND HEADWATERS

HYDROLOGICAL PROBLEMS AND ENVIRONMENTAL MANAGEMENT IN HIGHLANDS AND HEADWATERS:

Updating the Proceedings of the
First and Second
International Conferences
on Headwater Control

Edited by Josef Křeček,
G.S. Rajwar and Martin J. Haigh

A.A. BALKEMA/ROTTERDAM/BROOKFIELD/1996

ISBN 90 5410 726 X

Distributed in USA and Canada by: A.A. Balkema Publishers, Old Post Road, Brookfield, VT 05036, USA.

Preface

The International Conferences on Headwater Control are federal meetings which, since 1989, have aimed to improve communication across the national, occupational and disciplinary barriers that divide those concerned with environmental management and reconstruction in highland and headwater regions. These meetings have tried 'to unite the perspectives of the scientific researcher, the environmental practitioner, and the policy maker in a search for improved strategies for the reconstruction of threatened environments, livelihoods and communities in headwater regions'.

In the process, the meetings have created a near unique, informal, multidisciplinary, network which continues to link environmental managers and research workers from several disciplines in a score or so nations. These meetings still provide the only regular venue for the formal exchange of experience between some major international organisations devoted to different aspects of environmental management in the headwater and highland peripheries of different nations. These include: the International Union of Forest Research Organisations—especially the Forest Hydrology Group, the World Association of Soil and Water Conservation, which is especially concerned with erosion control in agricultural settings, the International Erosion Control Association, which is mainly concerned with erosion control in non-agricultural settings, the Earthwatch Research Corps, which is devoted to promoting field research in environmental conservation and promoting communication between the scientific community and the general public, and the Food and Agriculture Organisation of the United Nations, European Forestry Commission, which has created a Working Party on the Management of Mountain Watersheds.

The Headwater Control movement, founded by Dr. Josef Křeček in the Czech Republic, has also benefitted greatly from the sponsorship of the Czech, formerly Czechoslovakian, Scientific and Technical Society's Water Management Group, the Czech Forestry Society, the ENCO Environmental Consultancy of Prague, and the Agricultural University of Prague. It has also gained from the personal contributions of Dr. Martin Price, University of Oxford, England, and Dr. R.B. Singh, Delhi School of Economics, India, who have both undertaken major roles in the organisation of meetings.

The major themes which have been tackled by the Headwater Control meetings have included: environmental monitoring, hydrological management, erosion control, evaluation of the impacts of development, promoting better land husbandry and work towards the environmental reconstruction of headwater and highland environments. Contributions have concerned developments in science, technology, policy and implementation. The overarching concern has been to promote the self-sustainable development of headwater regions and the prevention of adverse environmental impacts both within these regions and in those downstream.

This is the fifth major publication arising from the first two Headwater Control meetings. The raw material of the conference deliberations appeared in two sets of conference proceedings. These were produced, in the main, to facilitate discussion at the conferences themselves. They were produced in small numbers and have, for some time, proved hard to obtain.

Křeček, J., Grip, H., Haigh, M.J. and Hočevar, A. (eds) 1989: *Headwater Control:* Volume 1/Volume 2. Plsen, Czechoslovakia: WASWC/IUFRO/CSVTS (ISBN 80-02-99929-0): 389 pp.

Křeček, J. and Haigh, M.J. (eds) 1992: *Environmental Regeneration in Headwaters (Proceedings of the Second International Conference on Headwater Control).* The Prague: Enco: 368 pp.

Subsequently, two collections of revised, selected, papers from the conferences have been prepared. These collected contributions from the first conference's special workshop on the interactions between forest removal and hydrological processes, and contributions from a special session at the second conference which examined the potential of GIS for the evaluation of environmental change in mountains.

Haigh, M.J. and Křeček, J. (eds). 1991: Headwater Management: Special Feature. *Land use Policy* 8(3): 171–205 (ISSN 0268-8377).

Price, M.J. and Heywood, I. (eds) 1994: *Mountain Environments and Geographic Information Systems.* London: Taylor and Francis: 309 pp. (ISBN 0-748-40088-5).

This volume completes the set. It includes formally revised and updated versions of some of the best of the papers which remain to be published from the general proceedings of the first two meetings. It also includes updated papers and reports of recent progress in projects from 15 of the nations represented at the meetings. These papers focus on technique, policy, process, strategy, and preliminary results. Relatively few involve the formal presentation of final results—a task more properly assigned to academic journals. However, these papers present the cutting edge of work in progress and some current thinking about process, methodology, and policy matters, which we hope, will be of immediate interest to those currently active in environmental management in headwater regions.

This volume has a topical focus which runs from geoecological problems—studies of the interactions between climate, soils, sediments, forests, agriculture and runoff, to land management issues in headwater and highland regions. These papers are prepared in celebration of the Third International Conference on Headwater Control: "Sustainable Reconstruction of Highland and Headwater Regions", held in Delhi, October, 1995. This international conference provided a platform for the discussions on the monitoring and conservation of highlands and headwater regions of the world. The problems and related hazards in the headwater regions were reviewed thoroughly to decide strategies for their sustainable reconstruction. The impacts include the erosion of soil, sediment deposition, deforestation, global climate change, acid atmospheric deposition, drying up of upland streams, etc. The conference also discussed the problems of Himalayan headwater regions which are stressed by soil erosion, landslides, opencast mining, deforestation and changes in vegetation cover.

We hope that, through the resources of India's Oxford and IBH, New Delhi, and their European partners, A.A. Balkema, Rotterdam, the Netherlands, the ideals and actions espoused by the headwater control network will interest, a still wider, international audience.

October, 1995   MARTIN J. HAIGH
                                                                                              and
                                                                                  G.S. RAJWAR

The Contributors

A. Hočevar, University of Ljubljana, Department of Agronomy Ljubljana, Slovenia.

Ádám Kertész, Geographical Research Institute, Hungarian Academy of Sciences, H-1388 Budapest VI, Andrássy út 62, P.O. Box 64, Hungary.

Anatoly F. Mandych, Coastal Geosystems Laboratory, Institute of Geography, Russian Academy of Sciences, Staromonetny per. 29, 109017, Moscow, Russian Federation.

Chris Soulsby, National Rivers Authority (Northumbria Region), Elden House, Regent Centre, Gosforth, Newcastle Upon Tyne NE3 3UD, England.

Dénes Lóczy, Geographical Research Institute, Hungarian Academy of Sciences, H-1388 Budapest VI, Andrássy út 62, P.O. Box 64, Hungary.

Eero Kubin, Finnish Forest Research Institute, Unionkatu 40A, 00170 Helsinki, Finland.

G.S. Rajwar, Department of Botany, Government Post Graduate College, Rishikesh 249 201, U.P., India.

H. Schreier, Resource Management and Environmental Studies, University of British Columbia, Vancouver, B.C. V6T 1Z3, Canada.

I. Vertinsky, Forest Economics and Policy Analysis Unit, University of British Columbia, Vancouver, B.C. V6T 1Z3, Canada.

Jiang Tong, Nanjing Institute of Geography and Limnology, Chinese Academy of Sciences, Nanjing, China.

Josef Křeček, Institute of Applied Ecology, Agricultural University of Prague, The Prague, Czeck Republic.

Joseph J. Kerekes, Canadian Wildlife Service, Environment Canada, Bedford Institute of Oceanography, P.O. Box 1006, Dartmouth, N.S. B2Y 4A2, Canada.

L. Kajfež-Bogataj, Department of Agronomy, University of Ljubljana, Ljubljana, Slovenia.

L. Marchi, National Research Council—Research Institute for the Prevention of Hydrological and Geological Hazard (CNR-1RPI), Corso Stati Uniti 4, 35020 Padova, Italy.

Luiz Carlos Baldicero Molion, Departamento de Meteorologia, CCEN/UFAL C. Universitária, BR-104 km 14, 57.072-970 Maceió, Alagoas, Brazil.

M.A. Lenzi, Department of Land and Agro-Forest Environments, University of Padova, via Gradenigo 6, 35131, Padova, Italy.

Maria Luisa Paracchini, Inst. di Ing. Agraria, Università degli Studi, 20133 Milano, Italy.

Martin J. Haigh, Oxford Brookes University, Oxford OX3 0BP, England.

N. Pristov, University of Ljubljana, Department of Physics, Ljubljana, Slovenia.

P.R. Tecca, National Research Council—Research Institute for the Prevention of Hydrological and Geological Hazard (CNR-1RPI), Corso Stati Uniti 4, 35020 Padova, Italy.

Ramazan Saraci, Tirana Municipality, Tirana, Albania.

S. Brown, Resource Management and Environmental Studies, University of British Columbia, Vancouver, B.C. V6T 1Z3, Canada.

Stanimir C. Kostadinov, Faculty of Forestry, University in Belgrade, Kreza Viseslava 1, 11030 Belgrade, Yugoslavia.

Sten Folving, Joint Research Centre, European Commission, 21020 Ispra (VA), Italy.

T. Vrhovec, University of Ljubljana, Department of Physics, Ljubljana, Slovenia.

Ted L. Napier, Department of Agricultural Economics and Rural Sociology, The Ohio State University, 2120 Fyffe Road, Columbus, Ohio 43210, USA.

W.A. Thompson, Forest Economics and Policy Analysis Unit, University of British Columbia, Vancouver, B.C. V6T 1Z3, Canada.

Contents

Introduction to Hydrological Problems and Environmental Management in Highlands and Headwaters

Martin J. Haigh, Josef Křeček and G.S. Rajwar

This book is devoted to the search for environmental self-sufficiencies in highland and headwater regions. Its aim is one promoted by the founders of modern India who called upon communities to seek self-reliance and to develop ways of life which allow a harmonious, non-destructive balance with natural systems (Gandhi, 1969). The routes towards this condition may have been redefined by modern environmental managers, adapted both in technological and cultural terms, however, the search continues for systems of land husbandry, systems of environmental management, which are self-sustainable and which do not place a tax on the environment nor require subsidy from outside.

The problems of environmental management in highlands and headwaters are the problems of the periphery. Highland and headwater regions tend to lie at the margins of a nation's economic heartland. Paradoxically, while such regions are often promoted as icons of national identity, they are also often among a nation's least developed and most economically backward areas. They tend to be less densely peopled, remote, areas of wilderness and of steep unstable slopes. For these same reasons, highland and headwater regions often include the last great reserves of natural resources in a nation. Their contribution to a nation is often measured in terms of their resources of forest, pure water, minerals, wildlife, and for the tourism and leisure industries. The 'purity' of a natural environment in these 'unspoilt' areas often has a huge emotional significance for a nation which impacts even upon those who may spend no more than a few hours of their lives as tourists in such regions. However, frequently, these areas are also the last redoubt for minority nationalities and for cultures which are very different from those of the national mainstream. This can lead to problems of social and political unrest. The populations of these regions frequently feel that their well-being is being sacrificed for those of the 'foreigners' of the national core and this feeling is always resented.

Highland and headwater regions, then, lie on the front lines of economic exploitation for many nations. They also tend to lie on the front-lines of political and cultural conflict. Since, by definition, highlands and headwaters are the source regions for a nation's water resources and since, normally, these marginal areas are the places where national boundaries meet, and sometimes overlap, highlands and headwaters have a significance which is far greater than either their internal resource potential or relatively small populations might warrant. Adverse changes in these areas can have very dramatic effects on both the environmental and political stability of a nation. Hydrological problems in the headwaters can change the patterns of flooding, sedimentation and river flow in areas hundreds of kilometres downstream. Problems of unrest on a nation's borders can send shock waves that rock governments thousands of kilometres away in the nation's capital.

In a way, headwater and highland regions present one of the greatest challenges to national administration. Superficially, they often appear economically backward and politically weak. They seem to be areas ripe for exploitation. However, they tend to be fragile in environmental terms and have an inherent political sensitivity (Haigh and Křeček, 1991). In sum, ill advised interference in such regions can have devastating environmental and political impacts (Johnson and Lewis, 1995).

These regions also present a great challenge for research scientists and those involved in environmental management. Highlands and headwaters are regions where modern environmental change may be very rapid. At the same time, they are places which have often been starved of investment in environmental research and monitoring, and places where research and environmental monitoring is often very difficult because of their isolated and/or complex terrain. Research institutes and universities are more often found in the large cities of a nation's heartland than in the fastness of its mountains or headwater peripheries. The Headwater Control meetings are devoted to overcoming this poverty of scientific and environmental data. The Headwater Control meetings are, genuinely, about the sharing of advice and experience between environmental scientists and managers working in the highland and headwater regions of the "Three Worlds". Almost uniquely, this group is neither an extension of American thinking nor is it the foster child of any United Nations agency.

Headwater regions also challenge the normal scientific practice which produces specialists. The problems of highland and headwater environments are problems of environmental sensitivity. They do not rest easily in any particular discipline —forestry, agriculture, hydrology, civil engineering, sociology, or politics. They are problems which require an integrated approach to the management of the land—the kind of integrated approach which today is associated with the phrase 'sustainable development', or more recently with the 'better land husbandry' movement.

However, the fact remains that there are few fora where environmental scientists from different disciplines routinely meet together, let alone meet in concert with planners, environmental practitioners, environmental managers and policy makers. The Headwater Control meetings remain almost the only international gathering where foresters, agriculturalists, environmental scientists and policy makers meet together as a matter of routine. Our federal conferences are promoted by a wide range of practically oriented international specialist associations, most of them non government organisations. Our meetings aim to unite the perspectives of the scientific researcher, environmental practitioner, and policy maker in a search for improved strategies for the management of threatened environments and livelihoods in highland, steepland and headwater regions.

The current major concerns of the Headwater Control conferences include:

- rebuilding the vitality of mountain environments,
- monitoring and reducing the environmental impacts of development in headwaters (especially the impacts of commercial forestry, tourism, road construction, mining, etc.),
- determining the characteristics of the hydrological regime in highlands and headwaters, especially with regard to the impacts of land use change, acid rain, climatic change, and the changing biological influences on the hydrological cycle,
- environmental monitoring in headwater environments and the collecting of the benchmark data which provide the basis for environmental action,
- erosion control and the management of steeplands water courses,
- the conservation and management of forest and water resources, and,
- the development and co-ordination of community action and community development for environmental improvement in mountains.

Aspects of all these activities may be found in the papers sampled for this book. Andrej Hočevar and L. Kajfež-Bogataj (Slovenia) examine the stability and dynamics of headwater ecosystems with regard to the potential of current climatic change. An introductory paper by G.S. Rajwar (India) highlights the challenges for environmental management in the Western Himalaya. Papers by Anatoly F. Mandych (Russian Federation) and Ramazan Saraci (Albania) produce benchmark analyses of hydrological process and sediment yield in headwater areas. Lenzi, Marchi, and Tecca (Italy) study the dramatic problems of debris flow dynamics in the Italian Alps.

Adám Kertész and Dénes Lóczy (Hungary) and Jiang Tong (China) look at the problems of agricultural soil erosion and conservation. Stanimir Kostadinov (Yugoslavia) examines the local impacts of land use change on headwater lands while Eero Kubin (Finland) focuses on the effect of clear cutting, waste wood collecting and site preparation on nitrate pollution of ground waters.

Luiz Molion tackles the global consequences of the massive deforestation affecting Amazonia, while Josef Křeček (Czech Republic) describes the regional consequences of acid atmospheric deposition on mountain watersheds in central Europe. Vrhovec, Pristov, and Hočevar (Slovenia) offer a theoretical model for the deposition of air pollutants in an Alpine headwater. Chris Soulsby (United Kingdom) examines the interactions between acid deposition and the hydrological controls on the leaching of aluminium in upland Wales while Joseph Kerekes (Canada) describes procedures for monitoring acid deposition in two headwater lakes in Nova Scotia.

Further contributions to the technical problems of environmental monitoring are offered by Hans Schreier and colleagues (Canada) and by Maria Luisa Paracchini and Sten Folving (Italy). Collectively they evaluate the use of GIS, DTM and remote sensing technologies.

The social and political aspects of environmental management in highlands and headwaters are examined by Martin Haigh (England), who compares the perceptions of different pressure groups lobbying for environmental action in the Himalayan environment, and by Ted Napier (USA), who reviews the problems of implementing environmental conservation practices in the farmlands of America's mid-west.

Three beliefs unite all of these works. This is the notion that the environment in headwater and highland environments is vulnerable to, and threatened by environmental change and human actions. The second is that something must be done to protect these environments. The third is that the solutions lie in the integrated environmental management.

Frequently, these ideas have been couched in terms of sustainable development. Sustainable development is defined as "development that meets the needs and aspirations of the present without compromising the ability of future generations to meet their own needs" (Brundtland, 1987:43). However, today, many will wonder if being sustainable is actually good enough?

In his review of the condition of the mountains in the Mediterranean region, McNeil (1992) shows how ecological and economic change may destroy the headwater environment and way of life. First in the Christian Mediterranean then in the Islamic nations, population overshot the carrying capacity of the headwater and highland regions. The result has been deforestation, soil erosion and population decline. Depopulation caused field abandonment, the collapse of agricultural terraces and the destruction of irrigation works. Pressures from external markets meant that distant demand could be focused with destructive effect. In case studies of villages in five mountain ranges: Taurus, Pindus, Lucanian Apenniares, Sierra Nevada/Alpujarra and the Rif, McNeil shows how the mountain way of life changed little before 1800. In the 19th Century, economic changes brought fleeting prosperity to some areas but this was based on easily depleted natural resources or an easily deflected trade route and it added popular warfare to the scourges of the environment. Massive deforestation between 1800 and 1950 was due to migration into the mountains. Consequent population pressure produced acute hardship and emigration by ecological refugees. The mountains carry the scars of environmental degradation from the distant history but more dramatically from recent times.

Looking forward, McNeil is pessimistic about the prospects for environmental protection in mountains and much impressed by their function as refuges for insurgency. He agrees with our conferences that societies which persist in such marginal and fragile headwater environments do so only through careful land husbandry. However, nowhere have mountain peoples devised systems of human ecology that have preserved a durable harmony or balance with nature. To date, only low population densities have preserved mountain environments (McNeil, 1992). This is no longer an option.

Certainly, being 'sustainable' is very much better than being unsustainable but this estate remains very far from ideal. Systems which are sustainable have to be sustained. This implies a conscious input, an act of additional and of sustained intervention. This point is more than an academic splitting of hairs. It has important implications.

The world is full of illustrations of sustainable land management in mountains, irrigation channels, agricultural terraces, and roadways (Pereira, 1989). Provided these engineered structures are sustained by repair and maintenance, they will last for a very long time. Equally, if they are neglected, for even a few years, they will fail. Terraces in northern Yemen and the Himalaya, the drains and channels of the reclaimed surface mine disturbed lands of Europe, all provide celebrated instances of the catastrophic land degradation which may result when maintenance is withdrawn.

Equally, there are buffer strips, grassed waterways, contour cultivation and conservation tillage practices which may be deployed for the sustainable cultivation of erosive soils and steep slopes. However, these practices create the impression that such sites can and should be cultivated. When the tradition remains, but the practices are forgotten, overlooked for a couple of years, or disturbed by civil instability, major problems can result.

So, sustainability is not enough. Ultimately, best practice land management must involve strategies and land uses which are self-sustaining. This means practices that do not require additional conscious intervention to sustain them and practices which, even if they are discontinued, will cause no additional penalty against the land. The best kind of land husbandry is that which minimises the necessity for conscious soil conservation.

Of course, no-one ever suggested that such a goal is universally attainable. Environmental managers are summoned, usually, in situations where the land is being badly managed. Soil conservationists are more often employed where soil has already been damaged and must be repaired than to protect soils which have yet to be damaged. Foresters and hydrological engineers more often find work in situations where societies wish to use land in a potentially damaging fashion than where they continue a tradition of living in harmony with nature. This situation is unlikely to change.

Nevertheless, these conferences and their supporters devote themselves to the search for a particular holy grail. This is the search for systems of environmental management and utilisation which are self-sustaining, and where nature and society coexist and coevolve in a harmonious balance. The fact remains that best practice environmental management, best practice land husbandry, is self-sustaining, not merely sustainable.

REFERENCES

Brundtland, G.H. 1987: *Our Common Future*. Oxford.

Gandhi, M.K. 1969: *The Voice of Truth*. Ahmedabad: Navajivan Trust.

Haigh, M.J. and Křeček, J. 1991: Headwater management: problems and policies—special feature. *Land Use Policy* 8(3): 191–205.

Johnson, D. and Lewis, L. 1995: *Land Degradation: Creation and Destruction*. Oxford: Blackwells.

McNeil, J.R. 1992: *The Mountains of the Mediterranean World: An Environmental History*. Cambridge University Press: Studies in Environment and History.

Pereira, H.C. 1989: *Policy and Practice in the Management of Mountain Watersheds*. London: Belhaven. Press.

Headwater Ecosystems: Their Stability, Productivity, System Dynamics and the Role of Climate Change

A. Hočevar and L. Kajfež-Bogataj

ABSTRACT

The impacts of agricultural and forest farm production are evaluated through an analysis of the flows of mass and energy into, and out of, the headwater ecosystem. The interactions between economic outputs from the system and inputs from air pollution and climate change are assessed in terms of their influence upon the resilience and stability of ecosystems.

Keywords: Headwater ecosystem, stability, system dynamics, climate change.

INTRODUCTION

Presenting an overview of headwater ecosystems studies regarding their stability and productivity is a very difficult task. This presentation will try to elucidate only a few of the headlines which one has to keep in mind.

First, we should answer a very important question: Why do we worry about headwater ecosystem studies? The answer is a very straightforward one: Its findings have great scientific value and great practical value, as well. Human managed ecosystems should be managed properly to ensure their stability in the long run, regardless of whether they are agricultural or forest, long existing or newly created, and taking into account their resilience to climate change. The aim must be to ensure optimum productivity at the lowest supporting input of mass and energy.

To be able to work towards these goals, first, we have to know how ecosystems function. Ecosystems must be studied as a whole using the system approach. Management should be based on physical, chemical and biological first principles with as little empiricism as possible, though studies should use all information available. On this basis, ecosystem functioning should be thoroughly examined such that as a final step it can be simulated on an appropriate time scale. This final model should be capable of recognising all the different parameters that define the system and its environment, and the different fluxes of mass and energy between them. Only with such knowledge and such tools we will be able to adjust the functioning of the ecosystem to obtain both stability and optimum productivity. In this work we will try to develop a theoretical basis for such a model and add some examples for agricultural and forest ecosystems.

THEORETICAL BASIS OF THE SYSTEM APPROACH METHOD

The first step in studying the ecosystem using the system approach is to define the system, its environment and fluxes of mass and energy between them (Fig. 1). The stability of the ecosystem from the physical point of view is defined by mass and energy balance on a proper time scale. This will be different for agricultural and forestry ecosystems and is related to rates of formation of organic matter and its mineralisation in the soil (Ulrich, 1986).

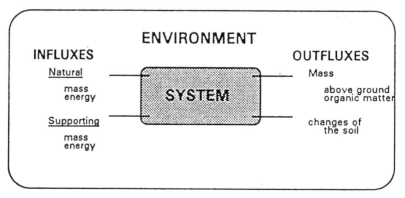

Fig. 1: The system and its environment.

A proper definition of the system is provided by Spedding (1988): "A system is a group of interacting components, operating together for a common purpose, capable of reacting as a whole to external stimuli: it is unaffected directly by its own outputs and has a specified boundary based on the inclusion of all significant feedback".

For further discussion it is very useful to mention a general definition of the agroecosystem, as well. Coleman and Hendrix (1988) define it thus: "An agrosystem is an ecosystem manipulated by frequent, marked anthropogenic modifications of its biotic and abiotic environments". This definition can be applied to most forest ecosystems in temperate zones, as well.

All studies of ecosystems are affected by the time and space scales. A very illustrative presentation of them together with atmospheric and biosphere processes is given by Sellers (1992), presented as in Fig. 2.

Tracing energy and mass flow through the plant canopy can be done by various dynamic models (Landsberg, 1981) which give the productivity of particular crop as a function of the environment. A very sophisticated example is described by Jorgensen (1991). This model takes into account many points of view and consists of various submodels as presented in Fig. 3. However, in our discussions, we will limit ourselves

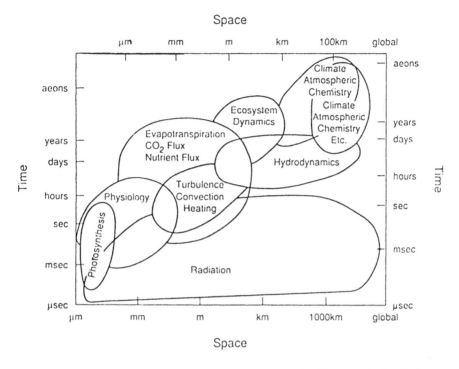

Fig. 2: Time-space scale diagrams of important processes as perceived by atmospheric scientists and biologists (after Sellers, 1992).

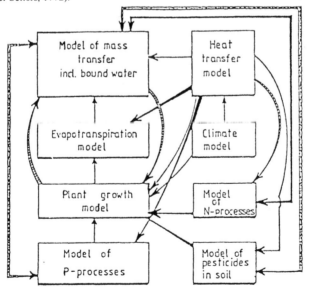

Fig. 3: Main features of an agricultural-environmental management model according to Jorgensen (1991).

to the first step, treat the headwater ecosystem as a black box and consider only inputs and outputs.

This system is defined as the headwater biosphere including pedosphere, surface boundary layer of the atmosphere and vegetation, including agricultural crops and forest. Its boundaries are: the relief, the upper boundary of plant canopy, the impermeable layer of the soil regarding water and solution fluxes, and the soil layer. Depth of about 1 metre, affected soil heat flux on daily time scale and 10 m regarding the yearly one (Fig. 4). To headwater ecosystem, in particular, the boundary, through which flows of water and other material move into the system and out of it, is rather sharply defined (Fig. 5).

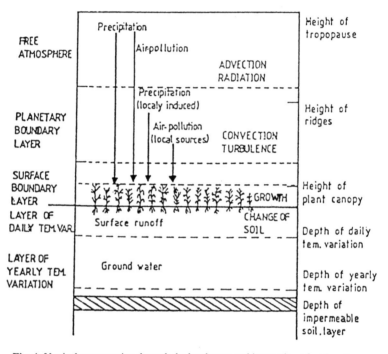

Fig. 4: Vertical cross section through the headwater and interaction of various layers.

Relief influences advection in the planetary boundary layer. So, the interaction between the atmosphere and the headwater ecosystem is rather complex. From the free atmosphere, besides precipitation, long range transport of air pollution also takes place. Relief boundaries influence the planetary boundary layer and its thermal stratification creating inversion layers and a specific spatial distribution of air pollution as illustrated by Gietl and Rall (1986) for the catchment "Grosse Ohe" and Vrhovec *et al.* (1992) for an alpine headwater in Slovenia. On the basis of the headwater topography, the mean emission potential for SO_2 can also be calculated as shown for Slovenian basins by

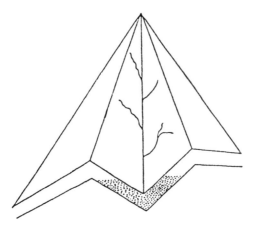

Fig. 5: Headwater ecosystem boundaries.

Petkovšek (1978). In sum, the surface boundary layer and plant canopy are influenced by both by the free atmosphere and by the planetary boundary layer (Fig. 4)!

In increasingly precise studies, the headwater biosphere can be divided into smaller elementary parts (Hočevar and Kajfež-Bogataj, 1986). Each may be characterised by different values of input and output parameters, of interrelations, of productivities and then summed to give figures for the whole headwater.

Fluxes of mass and energy between the system and its environment can be divided into natural and supporting ones added by man. The natural ones are determined by nature—radiation, advection, convection and conduction. The supporting ones are added to the system as work, fertilisers, pesticides and technology.

ENERGY FLUXES

The natural energy fluxes from the environment to the ecosystem, supplied by solar radiation during the vegetation period—April–September—are of the order of magnitude $30 \cdot 10^3$ GJ per hectare in middle Europe (Hočevar et al., 1982). This huge amount of solar energy influx keeps the temperature of the plant canopy and of the air at the level appropriate for biochemical processes in the plants and for evapotranspiration—which is essential for the process of photosynthesis. Only a very small part is used for photosynthesis itself (Fig. 6).

The support energy influxes differ greatly according to land use and vary with agricultural technology, as well. In Germany, Heyland and Solansky (1979) show how they also differ for various agricultural crops. However, its order of magnitude is about 18 GJ per hectare (Fig. 7). This is a figure which is three orders of magnitude lower than the natural energy input by solar radiation only! The main component of the support energy used in agriculture is fertilisers (Fig. 8 and Blankenhorn et al., 1978).

During the yearly cycle, the energy balance of the headwater ecosystem is mainly fulfilled. Only a few per cent of solar energy is stored as chemical energy and removed

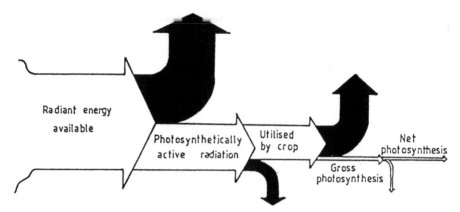

Fig. 6: Solar energy and its partition (after Spedding, 1988).

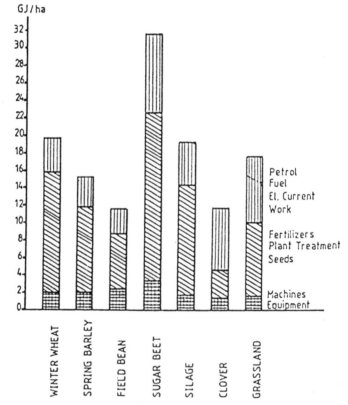

Fig. 7: Support energy input into crop production in Federal Republic of Germany (after Heyland and Solansky, 1979).

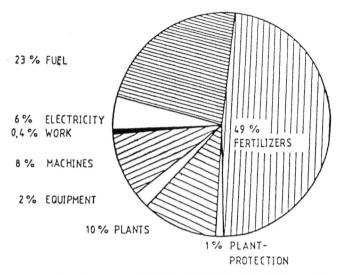

Fig. 8: Support energy input into crop production in Federal Republic of Germany (after Heyland and Solansky, 1979).

as crop. The majority is absorbed by the environment. This works as a good stabiliser taking up excess energy or covering the deficit if the differences are not too big: For instance, shortage of water can change the properties of plant canopy as albedo, leaf transmissivity, leaf area density, leaf area index and so on, and thus change the energy balance of the plant canopy.

The natural and supporting energy fluxes into the water ecosystem can be assessed rather precisely. Apart from the influence of climatic change, the situation is relatively straightforward.

MASS FLUXES

The situation with the fluxes of various masses into the headwater ecosystem is much more complicated. The water influx can be measured as precipitation. Fluxes out of the ecosystem evapotranspiration, run-off and groundwater can be estimated as in classical watershed hydrology. Some of this research work will be reported later in this section.

However, nowadays, there are other atmospheric inputs beside precipitation. The ecosystem is influenced by wet and dry deposits containing numerous pollutants and, at higher altitudes, by the interception of polluted fog water. For dry and wet deposits of sulphur, maps giving area distribution of their values were published for a large part of Europe more than ten years ago (NILU, 1977). The list of measured atmospheric pollutants and their deposits has extended during the last years, even in rural areas. For instance in catchment "Grosse Ohe" the mean annual depositions of ten elements and precipitation are recorded as given by Gietl and Rall (1986) (Table 1).

Table 1: Mean annual bulk deposition of various elements calculated from measurements in catchment "Grosse Ohe" between 25.7.1983 and 22.7.1985. Comparison of different sites in the open (OP) and under spruce forest (SF) (after Gietl and Rall, 1986)

Height (a. m. s. l.)		LOCATIONS			
		Teferlruck 769 m (OP)	Grueben 790 m (SF)	Hochgfeichtet 1250 m (OP) (SF)	1225 m
element	units				
H⁺	kgha⁻¹	0.6	1.1	0.7	2.4
SO_4	kgha⁻¹	35.5	59.3	44.5	109.5
NO_3–N	kgha⁻¹	6.9	9.9	8.9	19.3
Cl	kgha⁻¹	7.3	9.0	9.0	16.5
P total	kgha⁻¹	0.45	0.93	1.08	0.99
NH_4–N	kgha⁻¹	6.5	5.0	9.6	9.9
Ca	kgha⁻¹	3.9	10.7	5.6	21.0
K	kgha⁻¹	2.4	13.9	2.5	16.4
Na	kgha⁻¹	4.1	3.8	3.3	6.8
Al	kgha⁻¹	0.3	0.7	0.4	1.2
precipitation	mm	1105.5	830.1	1431.1	1315.5

For shorter periods, depositions of additional chemical elements including some heavy metals are known (Gietl and Rall, 1986). Similar orders of magnitude of deposition were found in the Solling mountains (Matzner, 1986) and near Vienna (Table 2).

Table 2: Representation of a forest landscape as a hierarchy (after Shugart and Urban, 1988)

Level	Boundary definition	Scale	Units
Landscape	Physiographic provinces; changes of land use or disturbance regime	10.000	ha
Headwater	Local drainage basins; topographic divides	100–1000	ha
Stand	Topographic position disturbance patches	1–10	ha
Gap	Large tree's influence	0.01–0.1	ha

In the evaluation of land use by agriculture or intense wood production, inputs of fertilisers and various materials for plant protection must be balanced with the output. Contemporary farming tries to operate in accordance with this postulate, though this ambition is never strictly fulfilled. For example, the several chemicals used for plant protection: pesticides, fungicides, insecticides, herbicides and so on, are designed to disintegrate in a short time period. Nevertheless, residues of these materials are left in the fields.

In a long time span, agriculture land use, which influences soil erosion and organic matter degradation, can reduce both soil fertility and depth of rooting of crops. "Ploughing in particular can cause erosion, and, over time, progressive mixing of surface and subsurface horizons, thereby altering water relations, soil organic matter status and nutrient availability" (Coleman and Hendrix, 1988) (Fig. 9).

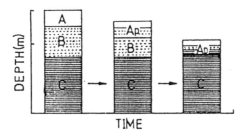

Fig. 9: Changes in soil profiles that result from simultaneous cultivation and erosion, with ultimate incorporation of subsurface material into the ploughing layer (after Schimel *et al.*, 1985).

The economically useful organic matter output of the headwater ecosystem is known as 'production'. For it numerous data are known. Data for the Federal Republic of Germany concerning various crops are presented on Fig. 10.

The balance of the material input and output the headwater ecosystem, and consequent changes of the ecosystem, are therefore hard to assess (Fig. 11). This is particularly true for the inputs coming "via atmosphere" known as pollutants. Also their pathway through the ecosystem to the soil and through plant roots to the plants is not yet fully understood. However, a lot of information about this problem can be found in the book entitled 'Air Pollution and Plant Metabolism' edited by Schulte-Hostede *et al.* (1988).

Till now, we discussed the energy and mass balance of the headwater ecosystem not mentioning the involved time scale. The extent and timing of the disturbances is what sets managed systems apart from the unmanaged ones. Human managed ecosystems are managed for net production (usually primary production), whereas most natural systems are much closer to a balance of production and respiration. Intermediate levels are reached in early succession stages.

Waring and Schlesinger (1985) illustrate the general relationships between the various components of forest metabolism regarding gross primary production from establishment to maturity (Fig. 12). Gross primary production is shown to peak after about 30 years, and net primary production some years before. This is a very important result for forest management from an economic point of view.

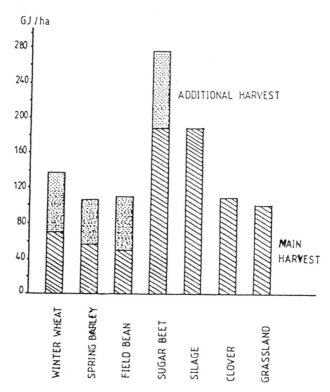

Fig. 10: Output energy contained in organic mass of various crops in Federal Republic of Germany (after Heyland and Solansky, 1979).

ECOSYSTEM STABILITY AND CLIMATE CHANGE

Examined from the perspective of different time scales, the headwater ecosystem shows different patterns of system stability and resilience. Resilience measures the speed at which a system recovers after a disturbance; $1/T_R$, where T_R is a recovery time. DeAngelis (1980) shows that system resilience can be related to two fundamental structural concepts: the energy flow through the system per unit of standing crop, and the recycling index that measures the mean number of cycles a unit of matter makes in the system before leaving it. A simple index which combines both of these, the transit time of a unit of matter through the system, provides a valuable index of resilience. The flux of energy or biomass through the system has an important effect on resilience. The higher the flux, the more quickly the effects of perturbation are swept from the system and so the higher the resilience (Pimm, 1988).

Regarding the space scale of the headwater ecosystem we cannot be very specific. According to Shugart and Urban (1988) a forested landscape can be thought of as a hierarchical system consisting of various levels (Table 2). One of these levels is

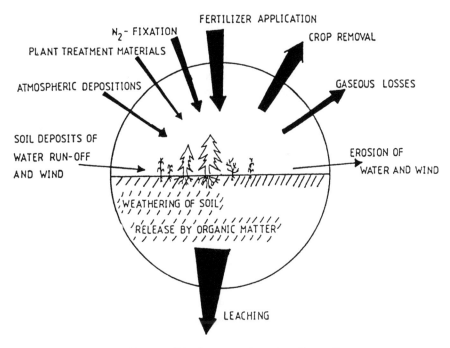

Fig. 11: Changes of biosphere status, in- and out-fluxes of masses.

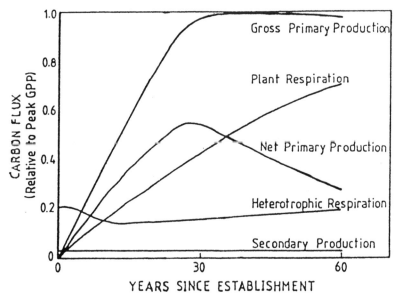

Fig. 12: Components of forest ecosystem metabolism as a function of time after Waring and Schlesinger (1985).

headwater. Landscape-level oriented studies have evaluated environmental parameters for soil and the surface air boundary layer. They give differences of their status as a function of the toposequence (Schimel *et al.*, 1985; Hočevar and Rakovec, 1983) and of the tillage system (Schimel *et al.*, 1985).

Conservation tillage practice greatly influences the output of water and nutrients from the headwater. Water run-off, soil transport, and total outputs of nitrogen and phosphorus were reduced by conservation tillage practices in an experimental watershed in USA (Langdale and Leonard, 1983). Interestingly, nitrogen and phosphorus concentrations were higher in run-off from the conservation tillage system, but a much lower total volume was exported!

Studies have been made for rather homogenous natural ecosystems of the whole headwater and for human managed ones of mixed land use. They both have specific peculiarities. According to Woods *et al.* (1983) large headwaters with mixed land use retain nutrients added as fertiliser to the cultivated system. Retention may be due to storage in the soil profile or to uptake by forest ecosystems downslope from the cultivated fields. Riparian ecosystems can accumulate a sizeable fraction of the nutrients lost from agroecosystems, thereby serving as nutrient "filters" for ground water and surface streams in agricultural areas (Lowrance *et al.*, 1984).

Ecosystem stability can be seriously disturbed by climate change. Climate is naturally subject to long-term changes and short-term fluctuations. Technological and scientific progress appears to accelerate climate warming by increasing the atmospheric concentration of radiatively active gases. Increasing concentration of greenhouse gases pose two different sorts of stresses on ecosystems (WMO, 1990). In addition to causing changes in climate, they can directly affect basic ecosystem processes, thereby influencing its health and composition. Primarily affected are biological and geomorphologic processes (e.g. photosynthesis and decomposition) which alter nutrient availability, reproduction and biomass productivity. Also very important are interactions between climate change and air pollution which can be particularly stressful to forest ecosystems. As air pollution increases, enhanced ozone concentrations and increased temperatures are expected to cause greater stress on plants and so may also damage watersheds.

Potential biophysical responses to climate change are not easy to determine because they can vary greatly from year to year and from region to region. Since there is great uncertainty about future temporal and spatial patterns of climate, the degree to which an ecosystem is sensitive to future climate change depends on the characteristics of the new temperature and precipitation regime, the characteristics of the site and of the crop or stand. Stands will be more sensitive where sites are moisture-limiting or temperature-limiting on plant survival and growth. Ecosystems will be more sensitive when they are in their establishment and late maturity phases or where they are composed of species that are at the limits of their natural ranges (Duinker, 1990).

The level of management in headwater ecosystem will play an important part in determining the kinds of effects future climate change might have, partly because we have to take into account the degree to which human intervention has already influenced ecosystem sensitivity to climate change. Examining biophysical changes on a regional scale is complex and adding humans as an additional variable further complicates the issue. Nevertheless, humans are the critical element in the study of ecological systems.

CONCLUSIONS

This paper has considered only those few points which we consider to be at very general importance to research. Nevertheless, all the studies have to merge into one picture although a very complex one. Only complex quantitative knowledge of headwater ecosystem function can give us the tools needed:
— to ensure the stability of the ecosystem taking into account its resilience, plus the undesirable effects of various pollutants, climate change and unsuitable land use,
— to ensure maximal economic productivity at the lowest possible supporting inputs of mass and energy which permit ecological balance.

We conclude our discussion with a sentence written by Rich (1988) who stressed the multidisciplinarity of ecosystem studies with the words: "Ecosystem ecology is no more (or less) a biological science than it is a chemical or physical science, because the subject is the "whole system", a unified, general and compelling scientific enterprise". Therefore, only scientists of various disciplines working together in close collaboration can be successful in understanding the ecosystem function and using the results in practice.

REFERENCES

Blankenhorn, P., Bowersox, T. and Murphy, W. 1978. Recoverable energy from the forest—an energy balance sheet. *Tappi,* 61 (4): 57–60.

Coleman, D.C. and Hendrix, P.F. 1988. Agroecosystems processes. In: *Concepts of Ecosystem Ecology.* Pomeroy, L.R. and Pomeroy, J.J. (eds.), Springer Verlag, New York, pp. 149–170.

DeAngelis, D.L. 1980. Energy flow, nutrient cycling and ecosystem resilience. *Ecology,* 56: 238–243.

Duinker, P.N. 1990. Climate change and forest management, policy and land use. *Land Use Policy* 7 (2). 124–137.

Gietl, G. and Rall, A.M. 1986. Bulk deposition into the catchment "Grosse Ohe". Results of neighbouring sites in the open and under spruce at different altitudes. In: *Atmospheric Pollutants in Forest Areas* (Georgii, H.W., editor). D. Reidel Publishing Company, Dordrecht, pp. 79–88.

Heyland, K.U. and Solansky, S. 1979. Energieeinsatz und Energieumsetzung in Bereich der Pflanzenproduktion. *Berichte ueber Landwirtschaft. Agrarwirtschaft und Energie*, 195. Sonderheft. Verlag Paul Parey, Hamburg and Berlin, 15–30.

Hočevar, A. and Kajfež-Bogataj, L. 1986. Assessment of ecosystem productivity in a watershed from meteorological point of view. III. Konference o biosfere. *Optimalizace Ekosystemu v Povodi.* ÈSVTS, Praga, 9–24.

Hočevar, A., Kajfež-Bogataj, L., Petkovšek, Z., Pristov, J., Rakovec, J., Roškar, J., Zupančič, B. 1982. Sončno obsevanje v Sloveniji—trajanje in energija. *Zb. Biotehniške fak. univ. E.K. v Ljubljani, kmetijstvo*, 96 pp.

Jorgensen, S.V. 1991. Environmental management modelling. In: *Introduction to Environmental Management* (Hansen, P.E. and Jorgensen, S.E., editors). Elsevier Science Publishers, B.V. Amsterdam, pp. 377–394.

Landsberg, J.J. 1981. The use of models in interpreting plant response to weather. In: *Plants and their Atmospheric Environment* (Grace, J., Ford, F.D. and Jarvis, P.G. editors). Blackwell Scientific Publications, Oxford, pp. 369–389.

Langdale, G.W. and Leonard, R.A. 1983. Nutrient and sediment losses associated with conventional and reduced-tillage agricultural practices. In: Lowrance, R., Todd, R.L., Asmussen, L.E. and Leonard, R.A. (eds). *Nutrient Cycling in Agricultural Ecosystem.* Coll. Agric. Spec. Publ. 23, University of Georgia, Georgia, pp. 457–468.

Leach, G. 1976. *Energy and Food Production.* IPC Science and Technology Press Ltd., Guildford, Surrey, England.

Lowrance, R., Todd, R.L. and Asmussen, L.E. 1984. Nutrient cycling in an agricultural watershed: I. Phreatic movement. *J. Environ Qual.* 13: 22–27.

Matzner, E. 1986. Deposition/canopy interaction in two forest ecosystems of northwest Germany. In: *Atmospheric Pollutants in Forest Areas* (Georgii, W.H., editor). D. Reidel Publishing Company, Dordrecht, pp. 247–262.

NILU. 1977. Long-range transport of air pollutants. Final report of a cooperat. tech. pro. for OECD, 300 pp.

Petkovšek, Z. 1978. Model for the evalution of mean emission potential of the air pollution with SO_2 in basins. *Arch. Met. Geoph. Biokl.*, Ser. B, 26: 199–206.

Pimm, S.L. 1988. Energy flow and trophic structure. In: *Concepts of Ecosystem Ecology* (Pomeroy, L.R., and J.J. Albert, editors). Springer Verlag, New York, 263–278.

Rich, P.H. 1988. The origin of ecosystems by means of subjective selection. In: *Concepts of Ecosystem Ecology* (Pomeroy, L.R. and Alberts, J.J., editors). Springer Verlag, New York, pp. 19–28.

Schimel, D.S., Coleman, D.C. and Horton, K.A. 1985. Soil organic matter dynamics in paired rangeland and cropland toposequences in North Dacota. *Geoderma* 36: 201–214.

Schulte-Hostede, S., Darrall, N.M., Blank, L.W. and Wellburn, A.R. (eds.). 1988. *Air Pollution and Plant Metabolism.* Elsevier Applied Science. London and New York, 381 pp.

Sellers, P.J. 1992. Biophysical models of land surface processes. In: Trenberth, K.E. (ed.). *Climate System Modelling.* Cambridge University Press, pp. 451–490.

Spedding, C.R.W. 1988. *An Introduction to Agricultural Systems.* Second edition. Elsevier Applied Science. London and New York, 189 pp.

Ulrich, B. 1986. Factors affecting the stability of temperate forest ecosystems. *18th IUFRO World Congress. Division. 1. Proceedings*, Ljubljana, pp. 121–135.

Vrhovec, T., Pristov, N. and Hočevar, A. 1992. Theoretical and model assessment of air pollution deposition in an alpine headwater. In: *Environmental Regeneration in Headwaters* (Křeček, J. and Haigh, M.J., editors), The Prague, pp. 278–287.

Waring, R.H. and Schlesinger, W.H. 1985. *Forest Ecosystems. Concepts and Management.* Academic Press, Orlando, Florida.

WMO, 1990. *Climate Change. The IPCC Impact Assessment.* AGPS Press, Canberra, 145 pp.

Woods, L.E., Todd, R.L., Leonard, R.A. and Asmussen, L.E. 1983. Nutrient cycling in a Southeastern United States watershed. In: *Nutrient Cycling in Agricultural Ecosystems* (Lowrance, R., Todd, R.L., Asmussen, L.E. and Leonard, R.A., editors). University of Georgia, Athens, Georgia, Col. Agric. Spec. Publ. 23, pp. 301–312.

Challenges for Environmental Management in the Headwaters of the Western Himalaya: An Introduction

G.S. Rajwar

ABSTRACT

This paper describes the challenges for environmental management in the headwaters of the Western Himalaya. Important causes and consequences of degradation in these headwaters have been explained. Forest clearing, fire, mining and other activities are causing deforestation and land degradation which result into threat to the survival of plant species and their habitats, low regeneration potential of some species, soil erosion and runoff, slope failures, landslides and plant succession problems. Proper measures for the environmental management of these headwaters are essential.

Keywords: Environmental management, forest degradation, mining, plant succession, Western Himalaya.

INTRODUCTION

Vegetation and mineral resources of the headwaters of the Western Himalaya have been utilised and exploited to a high extent for the last few decades. A number of activities have caused intense degradation of the ecosystems of the Western Himalaya. These include human as well as natural impacts. The important challenges resulting from these activities need to be solved for environmental management in the headwaters of this mountain system.

The Western Himalaya extends between the rivers Kali and Indus (west of Sutlej defile). Geographically, the Himalaya is divided into four zones: Sub-Himalayan zone, Lower Himalayan zone, Higher Himalayan zone, and the Tibetan or Tethys Himalayan zone. The Sub-Himalayan zone includes *duns* and Siwaliks. There are three major fold axes in the Himalaya: Himadri (Greater Himalaya), Himachal (Lesser Himalaya), and the Siwaliks (Outer Himalaya) (Wadia, 1953; Gansser, 1964). The Western Himalaya is divided into three regions in the east-west direction: the U.P. Himalaya (Kumaun and Garhwal Himalaya), the Himachal Himalaya, and the Kashmir Himalaya.

This paper describes important causes of degradation in the ecosystems of the Western Himalaya and challenges for the environmental management in this headwater region.

CAUSES OF ENVIRONMENTAL DEGRADATION AND CHALLENGES FOR ENVIRONMENTAL MANAGEMENT

The Himalaya represents one of the youngest mountain systems of the world; it is geologically unstable and made up mostly of stratified and sedimentary rock formation. The instability of the Himalaya is evidenced by the recent earthquakes. The important causes of degradation and challenges for environmental management of the headwaters of the Western Himalaya are given below.

1. FOREST DEGRADATION

Clearing of forests has been an important factor in the degradation of the headwaters of the Western Himalaya. Different forests in the montane and submontante zones are subjected to lopping and felling. "Chir" pine (*Pinus roxburghii*) and "Banj" oak (*Quercus leucotrichophora*) top the list of such trees. The clearing of forests is done for obtaining fuelwood and for construction purposes. The lopping pressure on chir pine trees is as high as 54% near the villages, whereas for half lopped trees it varies from 20 to 40% (Rajwar, 1989). The maximum loss of vegetation cover has taken place in various parts of Kumaun and Garhwal Himalaya. A number of workers from the Indian subcontinent and Western countries have observed and reviewed the environmental degradation in various parts of the Himalaya (Bahuguna, 1985; Myers, 1986; Singh *et al.*, 1984; Haigh, 1984; Ives, 1987; Hamilton, 1987; Ives and Messerli, 1989; Rajwar, 1988 b, 1989; Haigh *et al.*, 1990).

One of the important parameters of the analysis of deforestation is the index of diversity (Shannon and Wiener, 1963). In some parts of the Garhwal Himalaya, this index is low (not more than 0.84) for the forests under biotic stress (Rajwar, 1989). In a study on the vegetation of Bhagirathi valley, the density of saplings and seedlings has also been found to be lower for disturbed sites than that for the undisturbed ones (Rajwar, 1988b).

Intentional and unintentional forest fires which are common in different parts of the Western Himalaya also contribute much to the depletion of forest wealth. Fire sweeps and kills all the undergrowth including saplings and seedlings of plants. Pine forests are more susceptible to fire than oak forests. The fire thus severely affects the regeneration potential of the forest species.

A satellite imagery analysis of the U.P. Himalaya revealed that 37.5% of the area is actually covered by forest trees (Gupta, P.N., 1979), whereas Tiwari *et al.* (1986) recorded only 29% forest area for this region. The extension of agricultural area is also reducing the forest areas. In Kumaun Himalaya, the agricultural area is extending into the forest at an annual rate of 1.5%. The degradation of forests in the Himalaya causes more surface runoff and reduced infiltration (Haigh *et al.*, 1990).

The cattle population more than the carrying capacity leads to overgrazing of the pastures. In the Garhwal Himalaya, the cattle population is over 24 lakhs. These animals graze repeatedly the same pasture-land, and this activity harms the regeneration of many herbaceous species of plants. In many areas of the Western Himalaya such pasture-lands are not managed according to their carrying capacities.

The headwaters of the Western Himalaya are also rich in the medicinal plants which are exploited in large quantities for obtaining raw materials. The over-exploitation of such plants has reduced their natural populations, many of which are now in the endangered category (Jain and Sastry, 1980; Rajwar, 1981–82, 1982, 1984, 1989). Some of such endangered plants are *Rauvolfia serpentina*, *Aconitum heterophyllum*, *Berberis asiatica* and *Nardostachys grandiflora*.

2. MINING

Mining is another hazard to the environment of the Western Himalaya. The mining industry has shown a rapid growth across the nations. The total number of limestone mines in the Himalayan region is over 25% of the total number in India. A number of mines are under operation in the Western Himalaya. The U.P. Himalaya has the highest number of mines (61) followed by Himachal Pradesh (38) and Jammu and Kashmir (12) (Mineral Year Books, Indian Bureau of Mines). Limestone is extracted from all the headwaters of the Western Himalaya. Other minerals which are extracted from these headwaters include gypsum, graphite, bauxite and coal from Jammu and Kashmir, gypsum and rock salt from Himachal Pradesh, and dolomite, phosphorite and magnesite from U.P. Himalaya.

In the Western Himalaya, the mining in most of the cases is of contour strip opencast type. In this type of mining, the overburden is removed starting from the outcrop and proceeding along the hillside so that a series of benches is formed by successive cuts. The inside walls range generally within 10 metres, whereas on the outside debris is pushed downslope causing a great damage to the vegetation and streams.

The opencast mining activity in the headwaters of the Western Himalaya causes a serious damage to their ecosystems by removing the vegetation cover, soil and other minerals. In the Western Himalaya, the maximum land area affected by mining is occupied by Kumaun Himalaya (4820 ha) which is followed by Jammu and Kashmir (886 ha), and Himachal Pradesh (438 ha) (Negi, 1982). In the Garhwal Himalaya the mines are located in Dehra Dun-Mussoorie area and some parts of Tehri Garhwal.

Mining impacts on the mountains have been determined by Sahu (1988), Valdiya (1987, 1988) and others. Reports on the environmental studies in relation to geology and mining in the headwaters of the Western Himalaya are available in literature (Rajwar, 1981–82, 1986–87, 1988a, 1989, 1994; Valdiya, 1987; Negi, 1982; Bandyopadhyay *et al.*, 1983; Joshi *et al.*, 1988; Misra *et al.*, 1988; Sikka *et al.*, 1988).

The mines in the Western Himalaya are located in the montane and submontane zones. The forests are of *Pinus roxburghii* and *Quercus leucotrichophora* in the montane zone, and of evergreen and deciduous forests in the submontane zone. The excavated material slides down the slopes and removes the soil cover and vegetation resulting into slope degradation, soil erosion, runoff and lowering of the water table.

The dust accumulation on leaves is harmful to the physiology of plants. The mining activity has affected the epiphytic plants and other herbaceous species in addition to the loss of woody species and their habitats. The roads connecting the mining sites to the main roads also cause the degradation of the slopes and the forest areas. Mining explosions weaken the faults and rock formation which consequently accelerate the slope failures and landslides.

3. PLANT SUCCESSION PROBLEMS

The mining and other biotic activities responsible for deforestation and degradation of the vegetated areas result into the changes in soil composition and formation of wastelands which become unsuitable for the native plant species. The succession process in such areas is also affected and is invaded by xerophytic and other weeds. These species grow as colonisers in the degraded areas as they possess a high degree of tolerance to adverse soil and moisture conditions. Some of such plant species have relatively shallow root system which prevents the soil enrichment process, and thus, the succession would require a very long period to reach the climax stage of ecosystem development. The important exotic species colonising mined and degraded areas include *Euphorbia royleana, Lantana camara, Celosia argentea and Tagetes minuta.*

NEED OF A BALANCED POLICY FOR ENVIRONMENTAL MANAGEMENT

A balanced policy is required to meet the challenges for environmental management in the headwaters of the Western Himalaya. The reclamation of mined areas is the immediate need to revegetate them (Dadhwal *et al.*, 1985; Soni and Vasistha, 1986; Kilmartin and Haigh, 1988). Similarly, the areas degraded due to other causes require forestry plantations as suggested by Gupta, R.K. (1979) and Bradshaw and Chadwick (1980).

Mining should be restricted to a few mines in the Himalayan headwaters and that too not in the geologically fragile areas. Conservation of the endangered species and threatened habitats should be in the priority. Community forestry and other afforestation programmes should be based on local needs such as fuelwood, timber, fodder and fibre, and on the conditions of the environment. Native species should be preferred in the plantations as suggested by Chipko movement which has always piloted the need to plant native species and the development of community forests (Bandyopadhyay and Shiva, 1987). Chipko movement has played an important role in the conservation of the forests of the Western Himalaya.

REFERENCES

Bahuguna, S.L. 1985. People's response to ecological crisis in the hill areas. *In*: Bandyopadhyay, J., Jayal, N.D., Schoetti, U. and Singh, U. (eds.) *India's Environment: Crises and Responses*, Natraj Publishers, Dehra Dun, pp. 217–226.

Bandyopadhyay, J. and Shiva, V. 1987. Chipko: rekindling India's forest culture. *Ecologist* 17(1): 26–34.

Bandyopadhyay, J., Shiva, V., Ashis, G., Menon, A.G.K. and Nadir, K.L. 1983. *The Doon Valley Ecosystem Report on Natural Resources*. Centre for Study of Development Societies, Department of Environment, Govt. of India, New Delhi.

Bradshaw, M. and Chadwick, D. 1980. *The Restoration of Drastically Disturbed Land*. Blackwell Scientific Publications, Oxford.

Dadhwal, K.S., Katiyar, V.S. and Singh, D. 1985. *Eucalyptus* hybrid shows relatively better performance on mine spoil/debris soils. *Soil Conserv. Newsletter* 4(1): 9.

Gansser, A. 1964. *Geology of the Himalayas*. Interscience Publishers, London.

Gupta, P.N. 1979. *Afforestation, Integrated Watershed Management, Torrent Control and Land Use Development for U.P. Himalaya and Siwaliks*. Uttar Pradesh Forest Department, Lucknow.

Gupta, R.K. 1979. *Plants for Environmental Conservation*. Bishen Singh Mahendra Pal Singh, Dehra Dun.

Haigh, M.J. 1984. Deforestation and disaster in northern India. *Land Use Policy* 1(3): 187–198.

Haigh, M.J., Rawat, J.S. and Bisht, H.S. 1990. Hydrological impact of deforestation in the Central Himalaya. *International Association of Hydrological Sciences Publication* 190 (Hydrology of Mountainous areas): 419–433.

Hamilton, L.S. 1987. What are the impacts of Himalayan deforestation on the Ganges-Brahmaputra lowlands and delta? assumptions and facts. *Mount. Res. Dev.* 7(3): 256–263.

Ives, J.D. 1987. The Himalaya-Ganges problem: the theory of Himalayan environmental degradation, its validity and application challenged by recent research. *Mount. Res. Dev.* 7(3): 181–199.

Ives, J.D. and Messerli, B. 1989. *The Himalayan Dilemma: Reconciling Development and Conservation*. Routledge, New York.

Jain, S.K. and Sastry, A.R.K. 1980. *Threatened Plants of India: A State-of-the-Art Report*. Botanical Survey of India, Calcutta, and Department of Science and Technology, New Delhi.

Joshi, S.C., Pangtey, Y.P.S., Joshi, D.R. and Dani, D.D. 1988. Mining hazards in the Himalaya: a pilot study. *In*: Joshi, S.C. and Bhattacharya, G. (eds.) *Mining and Environment in India*, Himalayan Research Group, Nainital, pp. 286–293.

Kilmartin, M.P. and Haigh, M.J. 1988. Land reclamation policies and practices. *In*: Joshi, S.C. and Bhattacharya, G. (eds.) *Mining and Environment in India*, Himalayan Research Group, Nainital, pp. 441–467.

Misra, R.C., Mehrotra, R.C. and Joshi, M.N. 1988. Perception of environmental hazards in Dehra Dun valley due to exploitation of natural resources. *In*: Joshi, S.C. and Bhattacharya, G. (eds.) *Mining and Environment in India*, Himalayan Research Group, Nainital, pp. 299–312.

Myers, N. 1986. Environmental repercussions of deforestation in the Himalayas. *J. World For. Res. Manag.* 2: 63–72.

Negi, S.S. 1982. *Environmental Problems in the Himalaya*. Bishen Singh Mahendra Pal Singh, Dehra Dun.

Rajwar, G.S. 1981–82. Ecological problems of Mussoorie hills and their solution. *J. Himal. Stud. Reg. Dev.* 5-6: 73–76.

Rajwar, G.S. 1982. Endangered or rare plants of Garhwal and Kumaun Himalaya. *Himal. J. Sci.* 2: 38–40.

Rajwar, G.S. 1984. Exploitation of medicinal plants of Garhwal Himalaya. *Sci. & Environ.* 6: 37–41.

Rajwar, G.S. 1986–87. Mining and the Himalayan environment. *Everyman's Sci.* 21: 175–177.

Rajwar, G.S. 1988a. Limestone quarrying—a hazard to Mussoorie mountains and their ecosystem. *In*: Joshi, S.C. and Bhattacharya, G. (eds.) *Mining and Environment in India*, Himalayan Research Group, Nainital, pp. 294–298.

Rajwar, G.S. 1988b. The forest vegetation on the east and west facing slopes of the Bhagirathi Valley near Uttarkashi, Garhwal Himalaya. *In*: Prakash, R., Negi, S.S. and Shiva, M.P. (eds.) *Advances in Forestry Research in India*, International Book Distributors, Dehra Dun, Vol. 1, pp. 199–206.

Rajwar, G.S. 1989. Human impact on the forests of Garhwal Himalayas. *In*: Prakash, R. (ed.) *Advances in Forestry Research in India*, International Book Distributors, Dehra Dun, Vol. 3, pp. 229–241.

Rajwar, G.S. 1994. Mining hazards to the vegetation of Indian Himalaya. *In*: Melkania, N.P. and Melkania, U. (eds.) *Himalaya: Issues and Responses* (in press).

Sahu, K.C. 1988. Environmental impact assessment of mineral exploitation. *In*: Joshi, S.C. and Bhattacharya, G. (eds.) *Mining and Environment in India*, Himalayan Research Group, Nainital, pp. 3–14.

Shannon, C.Z. and Wiener, W. 1963. *The Mathematical Theory of Communication*. University of Illinois Press, Urbana.

Sikka, B.K., Swarup, R. and Saraswat, S.P. 1988. Impact of limestone quarrying on village economy and environment—a study of U.P. hills. *In*: Joshi, S.C. and Bhattacharya, G. (eds.) *Mining and Environment in India*, Himalayan Research Group, Nainital, pp. 260–269.

Singh, J.S., Pandey, U. and Tiwari, A.K. 1984. Man and forest: a Central Himalayan case study. *Ambio* 13(2): 80–87.

Soni, P. and Vasistha, H.B. 1986. Reclamation of mine spoils for environmental amelioration. *Ind. For.* 112: 621–632.

Tiwari, A.K., Saxena, A.K. and Singh, J.S. 1986. Inventory of forest biomass for the Indian Central Himalaya. *In*: Singh, J.S. (ed.) *Environmental Regeneration in the Himalaya*, Gyanodaya Prakashan, Nainital, pp. 236–247.

Valdiya, K.S. 1987. *Environmental Geology: Indian Context*. Tata McGraw-Hill Publishing Co. Ltd., New Delhi.

Valdiya, K.S. 1988. Environmental impacts of mining activities. *In*: Joshi, S.C. and Bhattacharya, G. (eds.) *Mining and Environment in India*, Himalayan Research Group, Nainital, pp. 29–42.

Wadia, D.N. 1953. *The Geology of India*. Mac Millan, London.

Water Erosion and Sediment Yield in Mountain Areas: Natural Preconditions

Anatoly F. Mandych

ABSTRACT

Spatial variations of water erosion and sediment yield in mountains are predetermined by the hierarchical organisation of hydrological systems and their capacity for the self-regulation of erosion and sediment accumulation. The scale and individual character of erosion and sediment yields are determined by other site-specific factors. Among these, runoff variability plays a critical role in the time-dependent changes of processes under discussion. The processes are illustrated by examples from the Caucasus Mountains.

Keywords: Water erosion, sediment yield, self-regulation by hydrologic systems, spatial and temporal variability.

INTRODUCTION

Mountainous regions are high energy environments in which the destructive and transportational role of running waters are particularly pronounced. Water erosion and sediment removal are agents of the multitude of endogenous and exogenous processes which affect the peneplanation of mountain regions.

The particular dynamism of natural processes systems in mountains can prove hazardous to human society. Extreme events, such as catastrophic soil wash from arable lands, mudflows, flash floods and the deposition of water-borne debris, may strike rapidly and with disastrous consequences. As human development of mountain areas increases, the potential for damaging interactions with water erosion and sediments transported by rivers also increases. The natural rate of these processes may be enhanced by human activities such as forest clearing, grazing, road construction, etc. However, the scale of these impacts must be assessed against that of natural processes and their fluctuations.

WATER EROSION FEATURES

Erosion of soils and rock debris is one of the key processes of earth relief renovation. Hillslope and channel erosion are the major suppliers of loose material to flowing water and so exert a decisive influence on sediment yields.

On mountain slopes, water erosion proceeds in combination with several other denudation processes: weathering, debris falls, landslides, creep, solifluction, and so on. The rate of erosion depends, not just on the energy of the flowing waters, but also on the supply of material which may be eroded and removed by water.

In any specific setting, the erosion rate depends on a great number of factors. The role of each of these changes through time in response to changes in the hydrometeorological regime, tectonic forces, and seismic impacts. The driving forces which control erosion do not remain constant from place to place. In mountains, water erosion displays pronounced spatial as well as temporal variability. This poses particular difficulties for the description of erosion in such regions and for its management during economic development.

SPATIAL SCALES OF EROSION AND SEDIMENT GENERATION

Viewing a watershed as a system of different structures involving hierarchically organised landscape units, it is possible to separate out those parts of subsystem, where particular water erosion factors are most manifest. These parts of the watershed may be recognised as nested hydrologic systems of different dimension (Table 1). In this context, their principal functioning involves the mobilisation and removal of loose material.

The SHS, Simple Hydrological System, is smallest. This is defined as a part of the mountain slope in which vegetation, soil, geomorphology and other landscape features are functionally homogeneous. From the standpoint of water erosion, the SHS's governing features are its capacity for the primary generation of surface and soil flows. The latter is realised as the result of a number of relatively simple water exchange processes such as precipitation, snow-melt and thaw, water-inputs from upslope, infiltration, evapotranspiration etc. The probability of soil flow initiation and its yield depends on the topographic slope of the SHS and on the hydrophysical properties of the soils and underlying loose deposits.

The CHS, Cascade Hydrological System, or slope system, is a combination of two or more SHSs joined by water flows which are parallel to the earth's surface. Within the CHS's boundaries, the structure of the SHSs change in accordance with the topographic slope, soil type, and distribution of beneath-soil slope sediments. The functioning of the CHS is manifested, not only in water flow generation on the slope surface, but also through interactions between surface runoff and transported debris. Since these interactions are different on the upper, middle and lower parts of the slope

Table 1: Key factors of water erosion and sediment yield forming

Level of hierarchy	Name and natural analogue of hydrologic system	Processes—factors ·	Features—factors
I	Simple Hydrologic System (SHS); site on slope.	— Weathering rate of parent rocks.	— Waterholding capacity and infiltration capability of soils and slope deposits. — Site surface slope. — Vegetation type. — Lithology of parent rocks.
II	Cascade Hydrologic System (CHS); slope.	— Water flow rate on slope.	— Slope inclination. — Slope aspect. — Variety of SHSs on slope.
III	Watershed Unit Hydrologic System (WUS).	— Intensity of channel flow concentration.	— Ratio of different slope areas. — Watershed's configuration in plan. — Watershed's geology.
IV	Complex Watershed System (CWS).	— Runoff amount and regime — Tectonic activity.	— Features of river networks. — Ratio of different altitudinal belt's areas. — Carrying capacity of the major river and its main tributaries.

system, the role of the different SHSs within the slope system cascade, CHS, is not the same.

In upper slope SHSs, the belt of weaker erosion, both the erosivity and carrying capacity of the flowing waters are small. The main movements of sediment downhill are due to the relatively slow processes of mass movement, solifluction, and the work of gravity.

In middle slope SHSs, the belt of active erosion, water flows energetically to remove sediments and, by so doing, creates a positive feedback loop in the system of 'water-slope' deposits. Figure 1 describes this most general, permanent, link between running waters and the loose materials on a hillslope. These trends are reinforced by vegetation. The effects of roots on soil and on evapotranspiration usually coordinate with the inherent, self-maintaining, interactions of the 'water-slope deposit' as slope development progresses.

The key element in the WUS, the Watershed Unit Hydrologic System, is the channel network. Water moving through this accomplishes the erosional reworking of sediments contributed from the hillslopes and their ultimate removal from the watershed. The sediment removing efficiency of the WUS depends on the way in which water flows are concentrated within the catchment during rainstorms, and on the intensity and patterns of snow-melt, other factors being equal. Under these conditions,

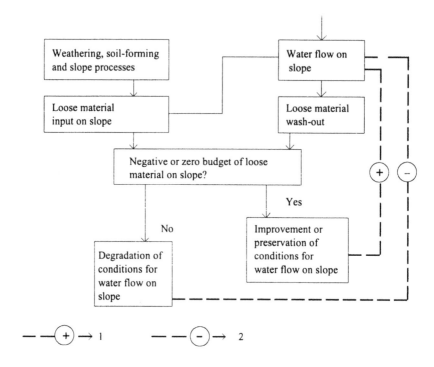

Fig. 1: Positive feedback between water flow and sediment removal in a Simple Hydrological System (SHS).

1—positive feedback, 2—negative feedback.

the key factors of sediment yield generation are slope angle and slope aspect, the ratio of different slope areas and the configuration of the watershed in plan.

The sediment yield of large rivers is dependent on all denudational processes developed in different parts of the Complex Watershed System (CWS). It also depends on the sediment transport capacities of the waters moving through the river network. In most cases, these conditions change regularly in the different altitudinal landscape belts that characterise mountains, while water runoff increases and channel slopes decline towards the lowest landscape zones.

In spite of the large diversity of conditions affecting erosion and sediment generation in mountains, it is usual for there to be an increase in the specific rate of sediment transport from its headwaters to middle reaches and a decrease from the middle reach to the river mouth (Makkaveev *et al.*, 1968). The belt of greatest erosion usually occupies the mid-altitude belt of large mountain watersheds, where the combination of highest power water flows (the product of water discharge by and channel gradient), density of river networks, and other conditions are optimal for erosion and sediment transport by rivers. However, forest vegetation may not have a determining effect on the progress of these erosion processes in the middle altitudinal belt (Hamilton, 1992; Hmaladze, 1964; Lal, 1983; Makkaveev *et al.*, 1968).

TEMPORAL VARIABILITY OF EROSION AND SEDIMENT YIELD

Changes of soil moisture in time, seasonal and long-term dynamics of vegetation, irregularity of precipitation and of thawed water on the ground surface, the random actions of certain geomorphological processes and seismic activity, these form the complicated temporal regime which governs erosion and which combines both causal and periodic elements. Greatest erosion occurs at some optimum combination of favourable factors.

Due to difficulty of field measurement of erosion characteristics on small plots and especially on slopes, which demand the setting up special experimental investigations, erosion rate is usually estimated by sediment runoff amount at an outlet cross-section of river. In such an approach the erosion amount is underestimated because of sediment accumulation within river watersheds (Dedkov and Mozjerin, 1984). The underestimation is great for watersheds on the plains and much less for those in mountains.

It has been known that the sediment yield of mountain rivers is characterized by very large intrayear and interyear variability. This is induced by river flow fluctuation and erosion rate changes taking place in their watersheds in different points in time (Thornes, 1980).

The higher variation of sediment yield compared to river discharges is the reflection of the not-proportional change of water erosion and sediment yield factors (Makkaveev, 1984). This is fixed by nonlinear relationships between water and sediment discharges, as used widely in hydrologic analysis (Figs. 2 and 3). However, even, with the help of such relationships, it is not always possible to separate the contribution to sediment yield of river erosive and transporting capacities, from an erosion processes' contribution proper, going on in I–III type hydrologic systems and which is mainly responsible for sediment income into mountain rivers.

The preceding can be illustrated by some examples from the Western Caucasus mountain rivers. Figure 2 shows the relationship between long-term averaged monthly suspended sediment discharge and averaged monthly water discharge for the small Abanos-Ckaly River (catchment area: 4.2 sq. km) which flows to the Black Sea, north of Batumi city.

It can be seen from Figure 2 that, at the same values of a river discharge, sediment discharge in the second half of summer and in autumn is twice and more as great as its value in the remainder of the year. Similar intrayear distinctions of sediment discharge values, with some variations, occur for such large rivers of the Western Caucasus as Codori, Rioni and Chorokhi. This phenomenon is caused by the greater unevenness of flood flow in the second part of summer and in autumn that, in its turn, is associated

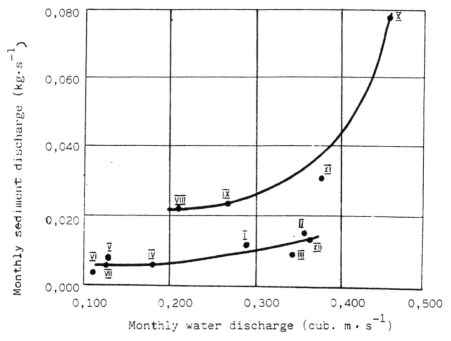

Fig. 2: Long-term averaged monthly suspended sediment discharge as a function of averaged monthly water discharge for Abanos-Ckaly River at Makhinjaury settlement.

with a predominance of storm rains within the precipitation regime of this period of year (Mandych, 1966, 1989). An increase of rainfall intensity stimulates water erosion on watershed slopes and is favourable for flash floods which have enhanced sediment transporting capacity.

Long-term fluctuations of sediment yield for the Codori River at Ganakhleba settlement is demonstrated on Fig. 3. This example reinforces the fact that sediment yield values at the same river discharge can vary by several times. Similar relationships of long-term yield change are typical for the other mountain rivers in the Western Caucasus region.

Such examples prove that intensive weathering of solid rocks in different altitudinal zones of mountain watersheds and high activity of other denudation processes in mountains generate large amounts of loose debris. This is the raw material for erosion. Its mobilisation depends on water flows occurring on the slopes and minor channel networks of watersheds. Runoff irregularity increases the erosive and transporting capacities of water flows to a large extent. Usually this factor is not adequately taken into account when evaluating the water erosion of mountain regions.

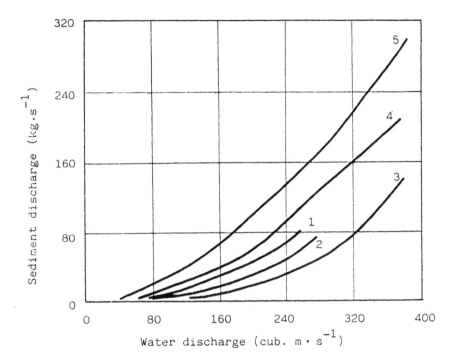

Fig. 3: Monthly sediment discharge as a function of monthly water discharge for Codory river at Ganakhleba village averaged over periods: 1—1934, 1935, 1947, 1949; 2—1952, 1953; 3—1954–1957; 4—1958–1960; 5—1961–1964.

CONCLUSION

The high spatial variability of water erosion and sediment yield in mountains is predetermined by the hierarchical organization of watershed systems. Inside the boundaries of small watersheds, the capacities of hydrologic systems interact to perform erosive work and sediment transport. The dynamic unity of erosion and accumulation processes manifests itself as a pattern of self-regulation.

Lithology of parent rocks, soil features, vegetation and other landscape factors impart a regional originality to the processes under consideration. In addition, they define the distinctions in erosion rate and sediment yield values between different regions.

Temporal variability of water erosion and sediment yield are predetermined by climatic features which induce variations in the formation of water flow patterns. Irregularity of slope wash and river flow is of paramount importance, increasing drastically the erosion rate and sediment yield, other factors being equal.

Economic activity on the territory of mountain watersheds has to be adapted to the space-temporal patterns of erosion and sediment yield. The role of these processes and their effect on population and economy is ambiguous. They can only realistically be assessed in terms of the larger system of the "mountains—adjacent plains". For example, the sediment yield decreases in Caucasian rivers owing to hydroelectric power stations construction has induced sediment deficits in the coastal zones of the Caspian and Black seas. This, in its turn, is one of the key factors causing severe erosion of these seas' coasts.

REFERENCES

Dedkov, A.P. and Mozjerin, V.I. 1984. *Erosion and Sediment Yield on the Earth*. Kazan University Publ. House, Kazan (In Russian).

Hamilton, L.S. 1992. The protective role of mountain forests. *Geojournal*, 27(1): 13–22.

Hmaladze, G.N. 1964. *Suspended Sediments of Rivers of the Armenian SSR*. Gidrometeoizdat, Leningrad (In Russian).

Lal, R. 1983. Soil erosion in humid tropics with particular reference to agricultural land development and soil management. In: Keller, R. (ed.), *Hydrology of Humid Tropical Regions*, pp. 221–239. IAHS Publ. No. 140, Wallingford.

Makkaveev, N.I. *et al.* 1968. Influence of uplifting relief development on bed river erosion and sediment yield in the Western Georgia. *Moscow University Bulletin*, 4: 52–58 (In Russian).

Makkaveev, N.I. 1984. General regularities of erosion and river channel processes. In: Makkaveev, N.I. and Chalow, R.S. (eds.), *Erosion Processes*, pp. 5–306, "Mysl" Publ. House, Moscow (In Russian).

Mandych, A.F. 1966. Influence of flood waves in mountain rivers on sediment discharge. *Meteorology and Hydrology*, 2: 21–28 (In Russian).

Mandych, A.F. 1989. Vegetation, water and climatic change. In: *Conference on Climate and Water*, Vol. 2, Publication of the Academy of Finland, 9/89, Helsinki, Finland, pp. 68–78.

Thornes, J.B. 1980. Erosional processes of running water and their spatial and temporal controls: a theoretical viewpoint. In: *Soil Erosion*. M.J. Kirkby and R.P.C. Morgan (eds.), Wiley-Interscience Publication, pp. 129–182.

Soil Erosion and Regional Ecological Management in the Headwater Area of Eastern China—A Case Study in the Tianmu Mountains

Jiang Tong

ABSTRACT

This paper provides a case study of soil erosion in the Tianmu mountains, which are the headwater area of the beautiful Taihu lake watershed in China. Since 1987, 5 auto-climate stations and 14 test-plots of soil erosion have been established in this area. The major features of soil erosion in the Tianmu mountains are as follows: (1) imbalance of soil material circulation, (2) decrease of soil thickness and increase of the coarse soil, (3) reduction of soil fertility, and (4) ecological environmental degeneration and frequent hazards. At last, improved strategies for soil erosion control are being implemented.

Keywords: Soil erosion, strategies of soil erosion control, headwater area, Tianmu mountains.

INTRODUCTION

The world is said to lose 5–7 million hectares of farmland each year to soil erosion damages. China has some of the most intensive soil erosion in the world. About 30% of the national territory suffers from severe erosion. The areas affected by the water and the wind erosion are 1.5 million and 1.3 million km², respectively. The amount of the eroded soil is 5000 million tons each year. Since 1949, the forest has been degraded severely, and the area of soil erosion has increased, day by day, with population growth and unscientific landuse, especially in the headwater area. The area of lost timberland is 17.3 million ha., about 17.36% of the total forest area. In the Yangtze River, for example, the area of soil erosion was 0.36 million km² in the 1950s, and 0.56 km² in the 1980s, about 1/3 of the total catchment area. The total amount of soil erosion is 2240 million tons each year, and the annual silt discharge is 500 million tons. Since the 1950s, about 45.5% of the lake area in the middle and lower regions of the Yangtze River has been lost including 35% of the water surface in the Don-ting Lake, and about 1210 million tons of sediment has deposited in the reservoirs. The area of desertification of China expands year by year. It increases its area by 1560 km² each year. Meanwhile, the agricultural ecological environment in some parts of China deteriorates. About 6.5 million ha. of land suffer from salinization. At present, about 1.3% of the cultivated area is low-productive land.

Tianmu mountains lie in the northwest part of Zhejiang province in the eastern coastal area of China, and they are also the junction of Zhejiang, Anhui and Jiangsu provinces. From southwest to northeast, they run about 180 km and have an area of 6,000 km². There are more than 60 mountain peaks over 1500 m, Longwang mountain, with an elevation of 1587.4 m is one of the highest peaks of the Tianmu mountains, and also one of the highest mountains of the eastern coastal area in the subtropical zone of China. Originating from the Tianmu mountains, and bringing 60–70% of total incoming water into Taihu Lake, the west and the east Tiaoxi rivers have created the flourishing Changjiang Delta area. They are the main water-supply and flood birthplace in Taihu Lake region (Fig. 1).

Located in the transition zone between the north and mid-subtropics, the climatic conditions of the Tianmu mountains are warm, humid, and affected by the south-east monsoon. The annual mean temperature is 8.9~16.0°C with the highest monthly mean temperature of 24.5~27.7°C and the lowest monthly mean temperature of –0.3~2.1°C. The active accumulated temperature ($\geq$ 10°C) is 2800~5150°C, lasting 199–266 days. The frost-free period is 140–240 days. The annual rainfall is as high as 1050~1850 mm, in which 2/3 of precipitation is observed during the plum rain and typhoon rain. The distribution of annual rainfall shows double peaks on the curve. The raindays ($\geq$ 0.1 mm/day) last 142~155 days. However, the annual precipitation has a non-uniform distribution.

Almost all the valley areas and steepland have been cultivated. The zonal forest, evergreen broad-leaved forest, remains in the upper part of the hills (650 m). In the mountains, the evergreen and deciduous broad-leaved mixed forest, and the deciduous broad-leaved forests are distributed between 600–800 m and > 800 m. Higher up are little mountain shrubs, mountain coppice and coniferous forest (> 1200 m, mountain summit).

The human activities mainly take place below 500 m. The valley plain and hilly area are densely populated. The population density and human activity decrease with increasing elevation. In the plain, steepland and low hills, the population density is 250~1000 persons/km². In the middle and low mountains, it is less than 250 persons/km². This area is a serious concern as increasing population pressure has caused slope land and forest to be brought under cultivation and into intensive soil erosion and soil loss.

Predatory mountain exploitation and undue farmland extension have caused the mountain's ecological environment to degenerate. Soil erosion has accelerated, and serious flood/drought disasters occur more frequently, especially in 1960, 1966, 1973 and 1980.

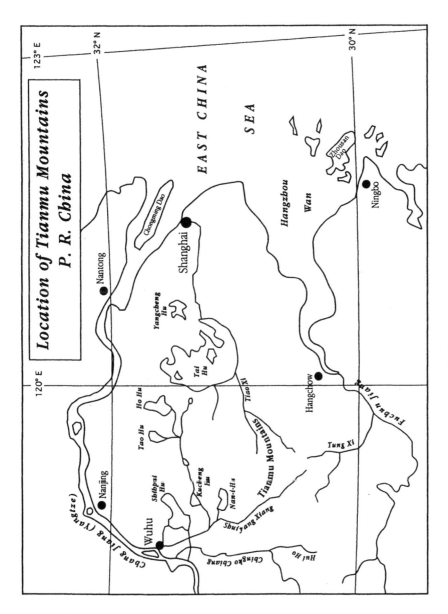

Fig. 1. The locations of the Tianmu mountains.

FIELD EXPERIMENTS

In 1987, a project sponsored by Volkswagenwerk Stiftung, Germany, and the Chinese Academy of Sciences was carried out. Five auto-meteorological observatories were established at different elevations along the west and the east Tiaoxi rivers. The locations and the major observation items are shown in Table 1.

Table 1: Field station observatories of landscape ecosystems in the Tianmu mountains

Station	Elevation (m)	Location	Observation
Station 1	1350	30°23′ 48″ N. 119°26′ 23″ E	P. T. WS. H.
Station 2	950	30°24′ 30″ N. 119°26′ 05″ E	P. T. WS. H.
Station 3	485	30°24′ 36″ N. 119°25′ 00″ E	P. Ra. WL. ST.
Station 4 (Fengchekou)	40	30°40′ 50″ N. 119°59′ 45″ E	P. Ra. WL. WT.
Station 5	20	30°48′ 19″ N. 120°03′ 46″ E	P. T. WS. H.

Note: P—Precipitation; T—Air temperature; H—Humidity; Ra—Radiation; WS—Wind speed; WL—Water level; WT—Water temperature; ST—Soil temperature.

As a contribution to research into the regulation of catastrophic soil losses, the integrated development, management and ecological reconstruction in such catchments, with the support of the National Natural Science Foundation of China, a small catchment, Fengchekou reservoir catchment, was chosen. This is located near Daixi town, 25 km south of Huzhou city, Zhejiang province.

The Fengchekou catchment is situated in the hills of the northeast slope of the Tianmu mountains, occupying an area of 1.69 km². A reservoir, with a capacity of 27.10 m², is at the catchment mouth. The highest peak in the catchment is Baijia peak with a height of 408 m. There are no permanent residents in the whole catchment. In 1988, 14 test plots were built in Fengchekou catchment on different slopes with landuses which correspond with the local natural and farming slopes (Table 2).

ANALYSIS OF SOIL EROSION

The Imbalance of Soil Material Circulation

The annual soil erosive modulus in the Tianmu mountains is 200–5000 t/km². Based on the Standard of Soil and Water Conservation in China, the acceptable erosive modulus is only 500 t/km². This means that the material removed from soil is more than that added into soil by natural processes. In Anji county, for instance, the eroded area, where annual soil erosive modulus is more than 500 t/km², accounts for 28.54% of the area (679.81 km²) of the whole county. The annual erosive amount is 34.25 × 10⁴ tons. About 11.9% of whole county area is in the moderately and intensively eroded area where the erosive modulus is more than 2000 t/km². The comparisons of erosive

Table 2: Test plots in Fengchekou catchment of the Tianmu mountains

Plot no.	Gradient	Orientation	Slope length (m)/ area (m)² Design	Actual	Land cover
1	27°	195°	10/50	10.3/51.5	Scrub community with sparse *Pinus massoniana*
2	27°	35°	10/50	10.2/50.8	Idem
3	9°	35°	10/50	10.3/51.7	Farmland with fruit trees (pear), pasture in 1989
4	22°	15°	20/100	20.3/102.3	Farmland with fruit trees (pear), fallow in 1989, water-melon and soybean in 1990
5	22°	25°	10/50	10.4/52.4	Idem
6	30°	190°	10/50	10.3/51.7	Tilted farmland with *Litsea cubeba*, fallow in 1989, 1990
7	25°	185°	10/50	10.2/50.8	Tilted farmland with *Litsea cubeba*, potato in 1989, fallow in 1990.
8	9°	90°	10/20	10.2/20.5	Farmland with bamboo (*Phyllastachys praecox*), fallow in 1989, pasture, corn and patato in 1990
9	9°	90°	5.10	5.2/10.4	Farmland with bamboo (*P. praecox*) pasture in 1989 (one season), pasture (two seasons), corn and potato in 1990.
10	4°	200°	10.20	10.2/20.5	Farmland, corn and water-melon in 1989, pasture and corn (one season) in 1990
11	4°	200°	5.10	5.2/10.6	Idem
12	28°	40°	20.100	19.9/99.3	China fir forest (13 years)
13	30°	20°	10.50	10.2/51.0	China fir forest (13 years)
14	24°	130°	10.50	10.7/53.5	Bamboo

Note: The slope length here is defined as the project slope length.

modulus and acceptable erosion on the north slope of the Tianmu mountains are shown in Table 3. This indicates that the area affected by the imbalance of soil material circulation is 8.4~28.5% of the whole county.

One case study found that nutrient storage of the withered trees and dead leaves on felled forest land (bare land) is only 49% of that on undisturbed forest land. Total chemical elements returned to the soil is 253 t/km² on the felled forest land, but 554 t/km² on the forest land.

The measurements in Fengchekou reservoir catchment show that different landuses are key factors affecting the soil loss. The soil losses from farmland are far greater than that of the forest land. Meanwhile, there is a close relationship between soil loss and the landuse intensity, and the soil loss of fallow land is less than that of the cultivated land (Compare Table 4 with Table 2).

Table 3: The soil erosive modulus and acceptable erosion on the north slope of the Tianmu mountains

County	E.m (t/km²)	A.e	E.m/A.e	Area of ≥ A.e in whole county
Anji	1765.5	500	3.5	28.5%
Deqing	1899.2	500	3.8	8.1%
Changxing	1149.2	500	2.3	11.4%
Huzhou suburbs	200.0	500	0.4	0

Note: E.m,: Erosive modulus; A.e: Acceptable soil erosion

Table 4: The erosive modulus (Aug. 1988–Aug. 1989) of the test plots in Fengchekou catchment of the Tianmu mountains

Plot no	1″	2″	3″	4″	5″	6″	7″	8″	9″	10″	11″	12″	13″	14″
E.m (t/km²)	–	–	78.0	80.3	150.3	45.8	61.6	108.3	247.5	78.8	111.3	10.9	4.3	110.7

Note: E.m,: Erosive modulus
There is a little material in the test plot 11″, 21″.

The Decrease of Soil Thickness and Increase of Coarse Soil

As soil is lost, soil thickness decreases, and the soil texture becomes worse. In Anji county, the area of soil with A horizon eroded is 18% of whole county area. In some parts of the Tianmu mountains, annual net removal of eroded soil is 3 mm.

The measurements in Fengchekou catchment indicate that surface soils become more coarse textured day by day due to the loss of the fine silts to erosion.

The Reduction of Soil Fertility

The availability of soil nutrients is governed by distribution down the soil profile. As soil is lost to erosion, the soil fertility steadily decreases. Based on the comparison of main soil nutritious materials, the organic matter in the forest land is 4 times more than that in the intensively eroded land, total N, total P are 4 times and 3 times, respectively.

In Fengchekou catchment, comparative analyses have been made between the fertility status of the undisturbed soil and the soil loss samples of 14 test plots. Measurements of organic matter, total nitrogen (%), total phosphorus (P_2O_5%), total potassium (K_2O%), available phosphorus (P_2O_3 ppm), and available potassium (K_2O ppm), indicate that the losses of the organic matter, total N, total P, total K vary with the different landuses. There are greater losses in farmland and lesser losses in the forest land. The amount of available P and available K, are greater in the samples of eroded soil than in the background soil samples for all landuses.

Ecological Degeneration and Environmental Hazards

Soil loss is the key factor in landscape ecological degeneration. Under the influences of human activities, the soil in the Tianmu mountains has become severely damaged and the mountain ecological balance destroyed. The tendency of landscape evolution is from scrub with *Pinus massoniana*, to waste land and barren land. As a result, flood hazard is aggravated.

In the West Tiaoxi River catchment of the Tianmu mountains, there are 233 records of heavy floods from the year 1363 to year 1949, with the maximum heavy flood in 1892. Since 1949, 11 heavy floods have occurred and the maximum is in 1963. During the 1963 flood, farmlands submerged occupied 91.8 km^2, and lots of sands, a volume of 1.0×10^8 m^2, was deposited in the reservoirs and river beds.

In Fengchekou catchment, measurements of transportation and deposition, suggest that erosive intensity has increased about 50% more than that in the 1970s. Only about 2.7% of the total soil losses in the catchment are taken out of the catchment through the Daixi River. The rest about 6–8% of the sediment is deposited in the reservoir. Lots of sediment, about 90% of total soil losses, remain in the catchment along the valley floors. This sedimentation increases the dangers of flood disasters in affected valley catchments.

STRATEGIES OF ECOLOGICAL MANAGEMENT

Human activity is the key factor in the catastrophic soil losses of the Tianmu mountains. Human activity leads to the soil erosion, and conversely, the soil erosion impacts on mankind. It is important that mankind also comes to the rescue of the damaged soils. With integrated research on the nature and law of soil erosion, the soil losses should be prevented and controlled.

Strategies to prevent soil erosion in the Tianmu mountains include soil loss reduction, anti-erodibility enhancement, and the mountain ecological balance recovery. The soil resources should be in rational utilization. Forbidding forest denudation and destruction strictly, combinative countermeasures should be adopted.

With comprehensive planning, prevention and control of soil erosion in the Tianmu mountains should become possible. The soil losses should be controlled in different natural regions with different methods. It is advocated that the problem concerning mountains, rivers, farmland, forests and roads should be tackled in a comprehensive way, at the scale of a small catchment ($\leq$ 10 km^2).

(a) In the area of less severe soil loss, the steeper mountains and hills (slope grade $\geq 20°$) should be closed to facilitate afforestation. Based on the different land suitable conditions, the mountains and hills (slope grade $\leq 20°$) should be planted to pastures, economic forests, fruits and crops and managed with a combination of landuse and soil conservation measures.

(b) In the area of moderate soil loss, caused by the combination of agricultural colonisation and forest exploitation of the mountain ecological system, forests should be recovered by the engineering practices and the biological engineering measures, such as the stripping and pit planting methods, band seeding, etc.

(c) In the intensive soil loss area, the soil losses should be controlled by field engineering and biological protective engineering, e.g., water storage dam, tree planting, etc.

CONCLUSION

This case study indicates the importance of research on soil erosion and the ecological protection in the headwater area. In the Tianmu mountains, the human activity is the key factor of soil erosion and landuse change is the direct cause. Because of soil erosion, land quality reduces and the flood disasters become more frequent. The soil loss in such headwater areas could affect strongly the socio-economic development of the lands downstream if the ecological environment is not improved in future.

In the past 5 years, the research on soil erosion control and the improvement of regional ecological environment has been put into practice. At present, some parts of soil erosion have been prevented and local ecological environments have been improved. The comprehensive exploit and management of the small catchment in the Tianmu mountains are carrying out extensively.

ACKNOWLEDGEMENTS

Authors are grateful to the National Natural Foundation of China and Volkswagenwerk Stiftung, Germany, for their financial support, and Prof. Dr. L. King, Geographical Institute of Justus, Liebig-University, Giessen, Germany, and Prof. Yu Xiaogan, Nanjing Institute of Geography & Limnology, Academia Sinica, for their kind guidance and help.

REFERENCES

Butterbach, K. 1988. *Klimatyologische Untersuchungen zur Erosivitat der Niderschlage im Einzugsgebiet des Tao Xi-Flusses, N-Zhejiang, VR China.-Diplomarbiet.* Geographisches Institut der Universitat Giessen, 121 pp.

Humi, H. 1983. Soil erosion and soil formation in agricultural ecosystems. *Mountain Research and Development* 3 (2): 131–142.

Jiang Tong, 1991. *Research on the Mountain Landscape in Huzhou City. Zhejiang Province, China.* Manuscript. 15 pp.

King, L., Yu Xiaogan and Jiang Tong. 1991. Wasser und Wind gefahrden die Landschaft. *Spiegel der Forschung* 8 (2): 6–13.

Kirkby, M.J. and Morgan, R.P.C. 1980. *Soil Erosion*, John Wiley & Sons Ltd., New York.

Nanjing Institute of Geography & Limnology, CAS, 1987. *Study on Long-range Prospect of Water-Land Resources and Agricultural Development in Taihu Lake Region.* Science Press, 184 pp.

Soil Erosion and Sediment Transport Depending on Land Use in the Watershed

Stanimir C. Kostadinov

ABSTRACT

The paper presents the results of erosion and sediment transport researches in 1980–1989. Three experimental watersheds were researched in the hilly-mountainous part of West Serbia (Yugoslavia). Parent rock and soil are the same in all three watersheds, but landuses are different. In the first watershed 40% total area is covered with well-stocked forests, ca 50% in the second one, and 70% in the third watershed. Such a distribution of forest cover is reflected in the intensity of erosion process in the watershed, as well as in sediment transport.

Keywords: Soil erosion, sediment transport, forests.

INTRODUCTION

Water erosion is a very complex process. The process starts at the moment when a rain drop hits the soil surface causing soil destruction. Detached and washed particles reach the hydrographic network of the stream and constitute its solid phase sediment. The development of erosion in a watershed depends on the sum of its basic interacting natural phenomena, such as: climate, relief, character and composition of parent rock and soil, state of vegetative cover, density of hydrographic network, etc. It is considered that permanent vegetation, first of all forest, is the strongest factor which opposes the adverse effects of erosion. Forests reduce soil erosion in the watershed, and reduce the yield of sediment from the watershed.

This paper presents the results of research into soil erosion and sediment transport in three experimental watersheds in West Serbia. These watersheds are characterized by similar natural features, the only difference being the degree of forest cover. In the Lonjinski Potok watershed, 70.35% of the area is under well-stocked forest, whereas the percentage in the Dubošnički Potok is 48.52%, and in the Djurinovac Potok watershed it is 39.50%.

METHODS

According to the aims of the investigation, in the selected experimental watersheds, discharge measurements were recorded, as well as suspended sediment and bedload transport, by means of regulated measuring sections. Precipitation data were also recorded. Water discharge was measured classically by means of measuring sections. Suspended sediment transport was measured by taking samples of water and determining the concentration of solids. Water sampling was repeated every day, and during the periods of flood wave, it was repeated several times a day. Bedload transport was measured by the volume method (Kostadinov, 1985). The sediment was retained in an accumulation space behind check dams. Bedload transport was measured after every flood wave determining the volume of the settled material by means of the precise level method.

RESULTS

Characteristics of Experimental Watersheds

The watersheds of the Dubošnički Potok and Djurinovac Potok are typical small watersheds which, together with their streams, make a characteristic hydrological-psammological system. All the three experimental watersheds are right-hand tributaries of the river Drina and they are torrential streams. The watersheds are situated on the territory of the community Ljubovija in West Serbia (Fig. 1). They are in a hilly-mountainous region, which can be seen from their main topographic features given in Table 1.

Table 1: Parameters of the watershed areas

Parameters	Symbol	Dubošnički Potok	Lonjinski Potok	Djurinovac Potok
Watershed area	F - km^2	1.2464	0.7656	0.544
Watershed perimeter	O - km	5.25	3.6	3.55
Watershed length	L_{sl} - km	2.48	1.4	1.4
Drainage density	G - km-km^{-2}	3.26	2.38	4.04
Local erosion basis	B_e - m	490.5	295.2	246.8
Mean elevation of the watershed area	N_{sr} - m	487.9	363.9	299.7
Mean altitudinal difference of the watershed area	D - m	243.4	159.1	106.5
Mean slope of the watershed area	J_{sr} - %	47.24	38.87	43.59
Stream-bed slope	J_1 - %	18.37	18.94	12.63
Erosion coefficient (Gavrilović)	Z	0.56	0.34	0.49

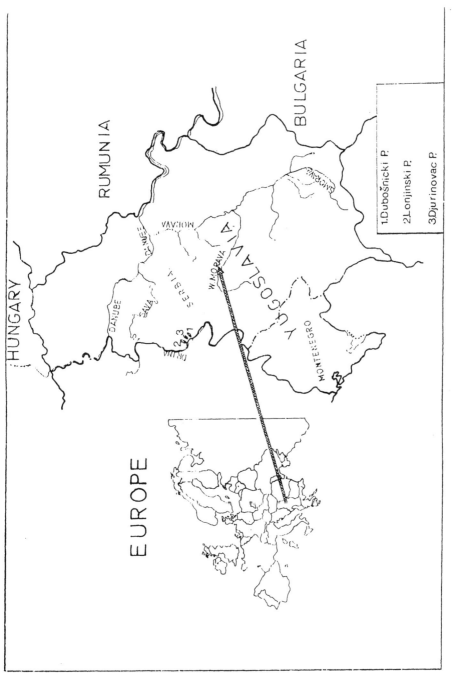

Fig. 1: The study area.

Parent rock is the same in all the three watersheds, i.e., sandy-schistose series consisting of metamorphosed sandstones, phylites, argilloschists and more rarely sericites, green schists, quartz breccias, quartzites and marbles.

The soil in all three watersheds is a skeletal acid brown soil.

The vegetative cover, i.e. land use, is different in the three experimental watersheds. Table 2 is a survey of land use. As can be seen, in the Lonjinski Potok watersheds, 70.35% area is under well-stocked forests, in the Dubošnički Potok watershed—48.52% and in the Djurinovac Potok watershed—39.5%.

Table 2: Landuse of the watersheds

Culture	Dubošnički Potok		Lonjinski Potok		Djurinovac Potok	
	km²	%	km²	%	km²	%
Pasture	0.0224	1.8	0.014	1.83		
Meadow	0.1768	14.18	0.07	9.14	0.0764	14.04
Plowland	0.0256	2.05	0.03	3.92	0.0136	2.5
Farm yard	0.0304	2.44	0.012	1.57	0.057	10.48
Orchard	0.0784	6.29	0.023	3	0.0168	3.1
Degraded forest	0.2848	22.86	0.078	10.19	0.1456	26.76
Well-stocked forest	0.6048	48.52	0.5386	70.35	0.2149	39.5
Bare land	0.0232	1.86			0.0197	3.62
Total	1.2464	100	0.7656	100	0.544	100

The vegetative cover in all three watersheds has the same characteristics, which is normal because the watersheds are near each other. Significant areas are covered by very degraded forests, transformed into very thin brushwood of eastern hornbeam, ash and oak, where very strong rill erosion occurs. The following associations are typical of the surviving well-stocked forests:

— Forest of Hungarian oak and bitter oak—*Quercetum farnetto-cerris*, Rud. L., with the sub-association *Carpinetum orientalis-serbicum*.

— Forest of beech in the mountainous belt—*Fagetum submontanum*. In addition, there are some artificially established plantations of black locust (*Robinia pseudoacacia*), Austrian pine (*Pinus nigra)* and Scots pine (*Pinus sylvestris)*. Their young plantings are sufficiently protected from erosion (Kostadinov, 1985).

Soil Erosion and Sediment Transport

As the result of the different degrees of forest cover, erosion processes of different intensities develop in the watersheds. Distribution and intensity of erosion processes in the watersheds, based on an erosion map plotted according to S. Gavrilović's method (Gavrilović, 1972), have been presented in Table 3. The most intensive processes of erosion occur in the Dubošnički Potok (Z = 0.56—medium erosion), then in the

Djurinovac Potok ($Z = 0.49$— medium erosion), whereas the weakest processes of erosion occur in the Lonjinski Potok ($Z = 0.34$—weak erosion); Z denotes the coefficient of erosion in a watershed or erosive area, according to S. Gavrilović.

Table 3: Distribution of erosion processes in the watersheds

Category	Z	Dubošnički P.		Lonjinski P.		Djurinovac P.	
		km²	%	km²	%	km²	%
I Excessive erosion	1.25	0.0862	6.92			0.048	8.8
II Intensive erosion	0.85	0.2296	18.42	0.0428	5.6	0.025	4.6
III Medium erosion	0.55	0.504	40.44	0.1922	25.1	0.204	37.5
IV Weak erosion	0.3	0.3698	29.66	0.3202	41.82	0.219	40.3
V Very weak erosion	0.1	0.0568	4.56	0.2104	27.48	0.048	8.8
Total		1.2464	100	0.7656	100	0.544	100
Mean coeff. of erosion		Z = 0.56		Z = 0.34		Z = 0.49	

Annual rainfall and annual characteristics of sediment transport are presented in Table 4.

DISCUSSION

The most intensive erosion is observed in the watershed of the Dubošnički Potok, then in the watershed of the Djurinovac Potok, and the weakest in the Lonjinski Potok. This is reflected by the mean value of the coefficient of erosion in all the three watersheds.

The situation of sediment transport is analogous to that of the development of erosion. The largest transport of sediment was observed in the Dubošnički Potok, and the smallest in the Lonjinski Potok.

The results of the research show that average specific annual module of total sediment transport in the Lonjinski Potok was 4.65 times smaller than in the Dubošnički Potok and 3.70 times smaller than in the Djurinovac Potok. If the value of mean annual sediment load in the Lonjinki Potok (Q_t—0.201 kg m³) is observed, it can be seen that it is 15.1 times lower than in the Dubošnički Potok and 6.20 times lower than in the Djurinovac Potok. This is the datum which illustrates even more drastically the state of erosion and sediment transport in these watersheds.

An important indicator of the sediment regime in a stream is the relation between bedload and total sediment. This is a very variable quantity which changes, even within the same stream, from the spring down to the mouth. It depends on many factors, first of all on the type of erosion processes and the transport capacity of the stream. As has already been pointed out, in the Lonjinski Potok there was no bedload at all. In the Dubošnički Potok the percentage of bedload in the total load amounted averagely to 53.9% (ranging between 0 and 85.3%). In the Djurinovac Potok (the least afforested

Table 4: Annual characteristics of sediment transport

Name of watershed	Year	H, mm	Sediment transport $m^3 km^{-2}$	$m^3 km^{-2}$	$m^3 km^{-2}$	r_t kg m^{-3}	$M_v M_g^{-1}$ %
Dubošnički	1980	1020.3	57.908	225.099	283.007	2.393	79.5
Potok	1981	984.5	81.548	30.913	112.461	0.463	27.5
	1982	794.8	76.462	34.859	111.321	0.839	31.3
	1983	687.3	28.263	14.69	42.953	1.384	34.2
	1984	705.5	254.22	52.451	306.671	1.15	17.1
	1985	509.5	44.334	256.646	300.98	7.477	85.3
	1986	722.6	32.669	33.763	66.432	0.987	50.8
	1987	873	316.006	464.556	780.562	10.922	59.5
	1988	602.1	3.308	0	3.308	0.077	0
	1989	747.1	106.985	57.99	164.975	4.678	35.1
Total		7646.7	1001.703	1170.967	2172.67	30.37	
Average Value		764.67	100.1703	117.0967	217.267	3.037	53.9
Lonjinski	1980	1054.7	16.032	0	16.032	0.089	0
Potok	1981	1011.2	38.013	0	38.013	0.13	0
	1982	779.6	48.158	0	48.158	0.216	0
	1983	768	40.263	0	40.263	0.272	0
	1984	906.1	64.313	0	64.313	0.186	0
	1985	591.3	13.463	0	13.463	0.074	0
	1986	612.2	3.012	0	3.012	0.042	0
	1987	995.5	119.844	0	119.844	0.386	0
	1988	737.1	90.985	0	90.985	0.464	0
	1989	875.2	32.628	0	32.628	0.154	0
Total		8330.9	466.711	0	466.711	2.013	0
Average Value		833.09	46.6711	0	46.6711	0.2013	0
Djurinovac	1981	1011.2	44.268	110.982	155.25	0.716	71.5
Potok	1982	779.6	101.467	76.827	178.294	0.93	43.1
	1983	734.1	10.783	0	10.783	0.078	0
	1984	906.1	24.643	18.833	43.476	0.238	43.3
	1985	591.3	11.868	22.316	34.184	0.613	65.3
	1986	703	30.853	63.528	94.381	1.243	67.3
	1987	674.6	129.445	263.25	392.695	1.008	67
	1988	889.1	152.379	295.456	447.835	6.716	66
	1989	916.6	72.747	123.343	196.09	0.942	62.9
Total		7205.6	578.453	974.535	1552.988	12.484	
Average Value		800.6222	64.2726	108.2817	172.5542	1.3871	61.7

H—annual rainfall in mm

M_r—annual specific transport of suspended sediment in m^3 km^{-2}

M_v—annual specific transport of bedload in m^3 km^{-2}

M_g—annual specific transport of total sediment in m^3 km^{-2}

r_t—mean annual sediment load in kg m^3

$M_v M_g^{-1}$—per cent of bedload transport.

watershed) the percentage of bedload in the total load amounted averagely to 61.7% (ranging between 0 and 71.5%).

Bearing in mind that the climate, parent rock and soil in all the three watersheds are the same, we shall analyze the natural phenomena as the causes of the above results. As for rainfall, the highest mean annual rainfall was measured in the Lonjinski Potok, and the lowest in the Dubošnički Potok (Table 4). Annual distribution of rainfall in all the three watersheds is similar. Maximum daily amounts were measured in May and in October.

As for the topographic features, the highest energetic potential is observed in the Dubošnički Potok, then the Lonjinski Potok, and the least in the Djurinovac Potok. Consequently, the Lonjinski Potok watershed, according to its rainfall and its energetic potential, could have the most intensive erosion and sediment transport, but the results of the research showed the opposite. The explanation is in the degree of forest cover.

The results show that forests have an important function in erosion control. The beneficial functions of forests are multiple. In a well-stocked forest, as a rule, there is no erosion even down the very steep slopes during heavy rains. The multiple erosion-control effect of the forest is the result of the protective impact of the crown and forest litter. Forest litter, owing to its high permeability and high water capacity enables the quick infiltration of rainfall. Forest litter enables the quick infiltration of rainfall. It is capable of absorbing 2–5 times more water than its weight in air-dry state. According to data by F.A. Gadziev (Zaslavski, 1983), the highest water capacity of forest litter reaches 400% its dry weight. The observations showed that, with the removal of forest litter, the runoff increased 5–10 times. The impact of the forest on erosion control depends on the choice of species, density of planting, age, grass cover, and forest litter (Zaslavski, 1983).

The effect of the forest on erosion processes is double: direct and indirect. The direct one is the protection of soil from the destructive power of rain drops. It also increases the resistance to the water which flows down the slopes, and in this way it reduces the erosive power and the capacity of water to carry sediment.

The indirect effect of the forest reflects itself in the good structure of the soil under permanent vegetation, which increases the resistance of the soil to erosion powers. The increased infiltration capacity of such soils leads to the reduced quantity and velocity of water movement down the slopes. Also, the specific fauna has the beneficial impact and it improves the structure and the infiltration capacity of the soil.

The data that in the Dubošnički Potok watershed there is a higher percentage of well-stocked forests compared to the Djurinovac Potok confirms the statement that only the watersheds with more than 70% area under a good vegetative cover can be considered as areas with erosion hazard reduced to a minimum (Kirkby and Morgan, 1984). As this condition is not fulfilled in the Dubošnički Potok, the higher energetic potential of the watershed (mean slope of the watershed, stream-bed slope, mean altitudinal difference of the watershed, and the local erosion basis) compared to the Djurinovac Potok watershed (Table 1) was the cause of the higher intensity of erosion, and also of sediment transport, when the other conditions were similar.

On the other hand, the Lonjinski Potok watershed also has a higher energetic potential than the Djurinovac Potok watershed, but owing to the fact that 70.35% of the watershed area is under a well-stocked forest, the erosion in the watershed is weak and, adequately, sediment transport was significantly lower than in the Djurinovac Potok.

It is interesting to note that similar results were recorded in a research by R.G. Tchagelishvili from The Republic of Georgia (Tchagelishvili, 1977).

CONCLUSIONS

The results of this research show that the degree of forest cover in hilly watersheds affects significantly the type and intensity of erosion processes, as well as sediment transport.

The highest intensity of erosion and the highest annual sediment transport was recorded in the Dubošnički Potok. Average annual specific transport of total sediment was M_g—217.267 m^3 km^{-2} (ranging between 3.308 and 780.562 m^3 km^{-2}). Average annual share of bedload in the Dubošnički Potok was 53.9%.

The Djurinovac Potok watershed is the next one where the intensities of erosion and sediment transport are compared. Average annual specific transport of total sediment was M_g—172.554 m^3 km^{-2} (ranging between 10.783 and 447.835 m^3 km^{-2}). Average annual share of bedload was 61.7%.

The lowest intensity of erosion and the lowest sediment transport was recorded in the Lonjinski Potok watershed with 70.35% area covered with well-stocked forests. Average annual specific transport of suspended (and total) sediment was M_r—46.671 m^3 km^{-2} (ranging between 3.012 and 119.844 m^3 km^{-2}). The transport of bedload was not recorded.

With the increase of forest cover up to 50%, the erosion control function of the forest does not increase to a significant degree. Further increase of forest up to 70% does change the impact of forest cover essentially and, in a hilly-mountainous region, this is reflected in the significant reduction of erosion. Sediment transport is also significantly reduced and there is no bedload transport. Such a degree of forest cover in a watershed dominates and forest cover has the greater effect than the energetic potential of the watershed (relief) and the erosion activity of rainfall, which has been proved by the example of the Lonjinski Potok watershed.

REFERENCES

Gavrilović, S. 1972. *Engineering of Torrent and Erosion*, "Izgradnja", Belgrade, Special edition, pp. 1–292 (in Seribian).
Kirkby, M.J. and Morgan, R.P.C. 1984. *Soil Erosion*. "Kolos", Moscow, pp. 1–415 (in Russian).
Kostadinov, S. 1985. *Research of Sediment Transport Regime in Torrents in Western and Southeastern Serbia*, Doctoral dissertation, Faculty of Forestry, Belgrade, pp. 1–314 (in Seribian).

Tchagelishvili, R.G. 1977. *Sediment Transport in the Small Watersheds with Different Degrees of Forest Cover in the Mountains of Georgia*, in *"Lesovedenie"*, N⁰ 5, Tbilisi, Georgia, pp. 72–77 (in Russian).
Zaslavski, M.N. 1983. *Soil Erosion Management*, "Visšaja škola", Moscow, pp. 1–319 (in Russian).

Sediment Transport in the Rivers of Central Albania

Ramazan Saraçi

ABSTRACT

The monitoring of runoff and sediment concentration at 15 stations on the three major rivers of central Albania since 1955 indicates that sediment concentrations range from 0.17 to 6.34 kg/m³. Reservoir sedimentation surveys suggest regional sediment yields ranging from 4.2 to 150 t/ha/year in watersheds between 66 and 15,600 ha. Although, there is some suggestion that sediment yield is affected by altitude, the main control of sediment release is the distribution or easily eroded rock types. The distribution of sediment concentration in the rivers of middle Albania is log-normal. Logarithmically transformed, the variables for river discharge and sediment concentration are linked by linear relationships. The relationship between annual water and sediment discharges is linear.

Keywords: River sediment loads, long term correlation between water and sediment discharge, denudation rates, central Albania.

INTRODUCTION

The three main rivers of the coastal lowlands of central Albania, the Ishmi, Erzeni, and Shkumbini rivers (Fig. 1) flow in parallel westerly courses, to the north and south of Tirana (41.20 N 19.49E), and drain watersheds totalling 4022 km². Although the major part of these rivers lies in the littoral lowland, their headwaters lie in the agricultural hill country of the Balkan uplands.

The Ishmi river in the north has its headwaters in the flysch hill country. This river, with its major tributaries, the Tirana, Terkuza and Zeza rivers, flows transversely across the limestones of the Dajti Anticline, before flowing over molassic and quaternary deposits to the sea.

The Erzeni river, in the heart of this province, flows in the main over the easily eroded sandy-clayey deposits of the flysch and molassic formations. Erodible and highly erodible deposits underlie about 76% of this watershed.

By contrast, in the south, the upper part of the Shkumbini river, as far as Murrash Gauging Station, crosses strata which relatively, much less erosive. Erosion resisting rocks cover 70–72% of the upper part of this basin, but only 50–60% of the lower basin where the more easily eroded clays and flysch deposits are found.

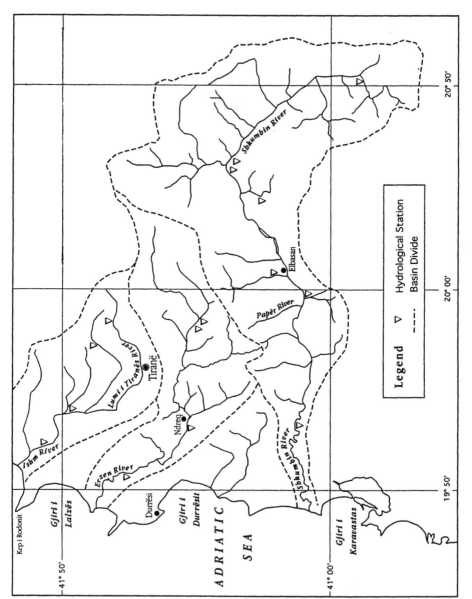

Fig. 1: Watersheds of Rivers in Middle Albania

DATA COLLECTION

Records of the suspended sediment concentration in these central Albanian rivers began in 1955. Fifteen recording stations have been active, including 5 which have collected records for 23–32 years, 9 for 11–18 years, and one for just 9 years. A summary of the data collected at these 15 stations is provided as Table 1. These data are supported by records of reservoir sedimentation which have been collected by the Water Directorate of Tirana and the Hydrometeorological Institute.

RESULTS

The records of sediment concentration indicate that sediment loads in the upper Ishmi river, which is dominated by very erodible strata, range between 1.99 and 2.35 kg/m³.

Those in the Erzeni watershed tend to be higher, ranging from 2.36 kg/m³ at Erzeni-Ibë Station up to 6.34 kg/m³, downstream, at Sallmanaj Station. However, sediment concentrations in the Shkumbini River, where the headwater geology is less erodible, are much lower, ranging from 0.427 to 3.17 kg/m³. The torrent Gustina in Fushë Bull, where only 8.2% of the land is underlain by the highly erodible land, is just 0.17 kg/m³. By contrast, the concentration is 3.48 kg/m³ in the Kusha torrent in Bradashesh, near Elbasan Town, where 85% of the watershed is underlain by the highly erodible strata. The pattern of variation since 1955 is displayed as Figure 2.

Early surveys of reservoir sedimentation preferred that rates of sediment yield that range from 4.2 to 102 t/ha/year in watersheds between 66 and 15,600 ha (Saraçi, 1974). More recent research suggests that sediment yields in this province range between 9.66 and 150 t/ha/year in watersheds of 65–1360 ha in area.

STATISTICAL ANALYSIS

The variation of sediment concentrations in the rivers of middle Albania may be modelled has a log-normal distribution (Fig. 3). This observation supports the earlier finding that the logarithms of average and annual sediment yields are normally distributed (Saraçi, 1989).

Correlation between records of river discharge and records of sediment concentration shows that, logarithmically transformed, these variables are linked by a strong (r: 0.7) linear logarithmic relationship (Fig. 4).

The double-mass curve relating annual water and sediment discharges is displayed is also approximately linear (Fig. 5). This suggests that human interference with these processes is not significant at this scale.

Each river basin is characterised by a particular relationship between specific annual sediment yield and the mean height of the watershed above sea level. This is influenced by geological variations and the distribution of the highly erodible rocks.

Table 1: Sediment loads at different stations

S. No.	Stations	Area of water shed F, km²	Mean height H, m	Sediment load				Area eroded rock in %
				Mean yield, Qs kg/s	Mean concen. S, kg/m³	Total yield, W tons 10³	Specific yield, q_s tons/ha	
1.	L. Tiranës-Zall-Dajt	70.8	911	6.65	2.35	209	29.5	54.8
2.	Gjole-Ura Gjol	468	404	29.5	2.03	929	19.9	71.6
3.	Ishmi-Sukth Vendas	651	367	45.9	2.26	1450	22.3	70.7
4.	Tërkuza-Pinar	113	662	7.80	1.99	246	21.8	66.9
5.	Erzeni-Ibë	247	819	19.5	2.36	614	24.9	58.1
6.	Erzeni-Ndroq	663	481	62.2	4.54	1960	29.6	75.0
7.	Erzeni-Sellmanaj	755	438	104	6.34	3280	43.4	74.0
8.	Përroi i Zallit Ibë	70.8	737	5.84	2.64	184	23.1	51.8
9.	Shkumbini-Sllabinjë	199	1260	2.22	0.427	69.9	3.51	30.3
10.	Shkumbini-Librazhd	851	1080	12.2	0.632	384	4.51	22.9
11.	Shkumbini-Murrash	1290	1050	38.5	1.09	1210	9.38	29.2
12.	Shkumbini-Papër	1960	891	106	1.96	3340	17.0	40.2
13.	Shkumbini-Rrogozhinë	2350	780	189	3.17	5950	25.3	46.9
14.	Përroi Gistimës Fushë-Bull	134	1200	1.05	0.17	33.1	2.47	8.2
15.	Përroi i Kushes Bradashesh	74.5	519	5.53	3.48	174	23.4	85.0

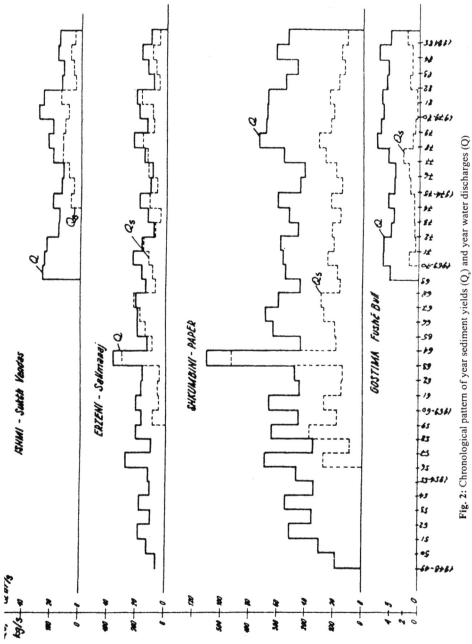

Fig. 2: Chronological pattern of year sediment yields (Q$_s$) and year water discharges (Q)

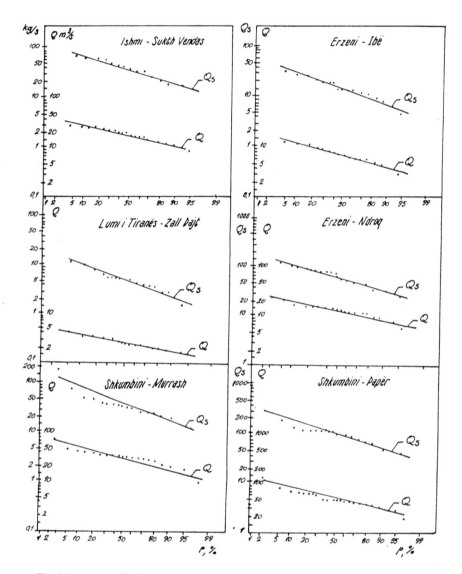

Fig. 3: Log-normal distribution of year sediment yields (Q_s) and year water discharges (Q)

In the case of the Ishmi river, there is a small positive association. However, for the Erzemi and Shkumbini rivers the relationship is negative. The gradient of the relationship is different for every watershed. Here, sediment yield increases with watershed area—which increases downstream.

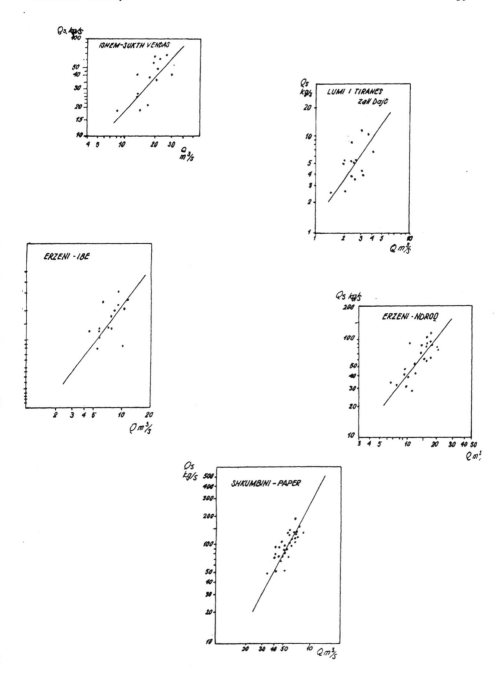

Fig. 4: Relations between year sediment yields (Q$_s$) and year water discharges (Q)

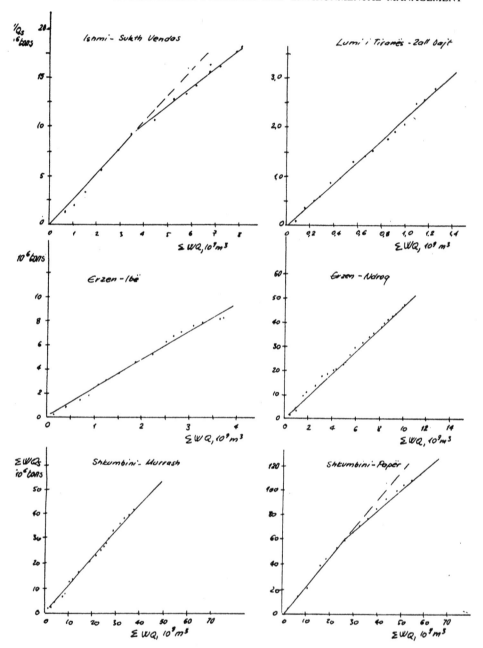

Fig. 5: Double-mass curves

CONCLUSIONS

The variation of sediment concentrations in the rivers of middle Albania may be modelled as a log-normal distribution. Log-linear relationships link records of river discharge and records of sediment concentration from individual rivers. A linear relationship links annual water and sediment discharges.

Sediment concentrations in the rivers of central Albania range from 0.17 to 6.34 kg/m^3. Sediment yields calculated from reservoir sedimentation surveys range from 4.2 to 150 t/ha/year in watersheds between 66 and 15,600 ha. Although, there is some suggestion that sediment yield is affected by altitude, the main control of sediment release is the distribution of easily eroded rock types.

REFERENCES

Saraçi, R. 1974. Sedimentation in reservoirs and debris basins in mountainous areas. *Met. and Hyd. Studies, Tirana* 6: 161–173.
Saraçi, R. 1989. On the correlation between the sediment load and water flow. *Met. and Hyd. Studies, Tirana* 13: 160–163.

Soil Erosion Control in Hungary

Ádám Kertész and Dénes Lóczy

ABSTRACT

About two-thirds of the total area of Hungary is agricultural land and about 25% (2297 ha) are affected by soil erosion processes. On hilly areas the effect of soil erosion is very significant causing enormous damages. Erosion hazard on hill landscapes is increased by the fact that a considerable part of the surface is mantled by loess and other unconsolidated sediments. Soil erosion control, agrotechnological, biological and technical measures, is a central task of farming on hillslopes.

This paper offers an overview of soil erosion control measures followed by a case study of the soil erosion processes of a small watershed. Soil loss assessment for different landuse types provides the basis for the design of control measures of each field within the watershed.

Keywords: Soil erosion, erosion hazard, erosion control, sediment yield, Hungary.

INTRODUCTION

Soil erosion studies are of great importance not only in surveying mere soil loss, but also informing about the sites where the removed soil accumulates together with its most common transport medium, i.e. water. Redeposited soil masses may contain chemicals (surplus amounts of fertilisers, pesticides or herbicides), which may cause environmental pollution. Considering water and soil together, we may arrive at a better understanding of geomorphological processes as well as at a better design of soil conservation measures and farming practices.

In Hungary erosion endangers a major natural wealth, fertile soils. More than one-third of agricultural land (2.3 million hectares) is affected by water erosion (13.2 per cent slightly, 13.6 per cent medium and 8.5 per cent severely eroded) and 1.5 million hectares by wind erosion (Stefanovits and Várallyay, 1992, see Table 1).

SOIL CONSERVATION IN HUNGARY

Erosion-control measures are usually implemented in a broader framework as one of the many tasks of soil amelioration. In K. Géczy's definition (in Szabó, 1977) *soil amelioration* means any influence on soils aimed at either increasing fertility in the long term, or eliminating or reducing adverse natural effects, including erosion, on

Table 1: Soil erosion in Hungary (calculations after Stefanovits and Várallyay, 1992)

	Thousand hectares	% of the total area	% of the agricultural land	% of the eroded land
Area of the country	9303	100	–	–
Area of agricultural land	6484	69.7	100	–
Arable land	4713	50.7	73.0	–
Total eroded land	2297	24.7	35.3	100
strongly	554	6.0	8.5	24.1
moderately	885	9.5	13.6	38.5
weakly	852	9.2	13.2	37.4

cultivation. Amelioration covers all those activities (agrotechnological, biological, chemical and technical) which are necessary to preserve the present state and fertility of the soil without causing adverse off-site effects.

Complex soil amelioration investments embrace the preservation and improvement of soil quality and fertility (Várallyay, 1986). The complex nature of the interventions is emphasized since, although single amelioration measures may be efficient, the combination of influences generally produces a better effect. The techniques employed may also serve several purposes. Soil-amelioration services, providing expertise on fertilization, soil utilization and the application of chemicals, are the day-to-day tasks of regional soil-conservation units.

The scope of soil amelioration in Hungary has undergone substantial changes over the last decades. At first, the primary emphasis lays in reducing the influence of factors restricting productivity, while in the recent past the main task was to achieve a greater security of production through minimizing the fluctuation of yields. The present concept of soil amelioration covers various interventions, from technical ones to land reclamation.

Previously, soil amelioration focused on erosion control, drainage, irrigation, chemical soil improvement and other local tasks. Later on there was an increasing trend to *concentrate* the resources in critical areas. The increasing importance of this *watershed amelioration* is reflected in the distribution of subsidies. In the 5th Five-Year Plan, only farm and other local amelioration projects existed. In the 6th Five-Year Plan (1981–6) 68 per cent of subsidies went to complex watershed amelioration schemes. The increase in yields after amelioration was 30 per cent on average in the 6th Plan, and the level of security was also higher (Varga, 1985).

Since 1989 reprivatisation of land has been going on and a new system of subsidisation has not yet been elaborated.

SOIL EROSION CONTROL IN HUNGARY

Soil erosion affects 2297000 ha of Hungary's land, 25 per cent of the country's total area (Stefanovits, 1977). Agricultural land makes up 70 per cent and about 35 per cent of it has a water or wind-erosion hazard. As a result of the post-war reorganization of agriculture, average field size has been multiplied and mechanization has also created a new situation for the preservation of soil fertility.

Soil conservation became part of the state agricultural policy (Stefanovits, 1977; Várallyay and Dezsény, 1979). In 1957 the National Soil Conservation Council (later renamed Amelioration Council) was formed to co-ordinate conservation activities and to identify priorities in this field.

Soil-conservation planning began at three levels:

1. Plans for the whole country or for selected regions or watersheds were made at 1:500 000 to 1:100 000 scales;
2. Smaller regions and partial watersheds were surveyed and recommendations were made at 1:50 000 to 1:25 000 scales;
3. For individual farms, scales of 1:25 000 to 1:10 000 were preferred.

The protection of agricultural land had been legislated for a long time by the 1961/ VI Act. It was partly the bureaucratic constraints on land ownership and partly wastefulness of land use in terms of the conversion of prime land for other purposes, and the resulting steady reduction of cropland, that necessitated comprehensive legislation of land.

The 1987/I Act (popularly called the Land Codex) includes provisions on the soil-conserving cultivation of land, according to its physical endowments and actual land use. Cultivated crops have to reduce soil loss and, if necessary, special measures also have to be applied. Cultivation must promote the balance of nutrients and preserve soil fertility. For any construction requiring land, tracts of poorer-quality land must be used. In general, land is now gradually acquiring commodity status in Hungary. This process is assisted by the new land-evaluation system, decreed in 1979 and now being introduced. With the exception of soil-conservation and irrigation projects, any other-purpose construction can only acquire land under permission and after paying a set contribution.

The paragraphs on landownership also have direct implications for soil conservation. In order to establish a more rational cultivation, farming units (state and co-operative farms) were free to exchange tracts of land between each other and to form fields better adjusted to the physical conditions. If no agreement was reached between the parties involved, a farm was allowed to apply for action at the land inventory office.

WATER EROSION

Its relief and drainage conditions make Hungary rather severely affected by water-erosion processes. In the mountain and hill regions surplus runoff, loss of soil, nutrients and fertilizers and the accumulation of washed-down material present problems over an area of 1300000 ha.

The territory of Hungary receives water from the encircling mountain ranges. Parallel with flood waves on rivers, groundwater levels rise in the lowlands, causing damage by excess water over 330000 ha of agricultural land. About 100 million tonnes of soil and 1500000 tonnes of humus are removed every year in runoff (Stefanovits, 1977).

Erosion control is a central task of farming on hillslopes. The techniques applied are usually grouped as either agrotechnological, biological or technical.

The most common practices of mechanical soil conservation are ridging (ridges are obliterated when ploughing on less than 12 per cent slopes and maintained on 12–17 per cent slopes) and terracing of 17–25 per cent slopes. Both types of erosion control are usually supplemented by grass waterways. Where runoff surges are particularly destructive, waterways are built of more durable materials.

Crop-rotation planning and agronomic techniques (e.g. subsoiling and ribbed rolling) are employed to promote the infiltration of rainwater into the soil.

WIND EROSION

Soil erosion by wind affects 16 per cent of Hungary's surface. Damage is primarily caused on sandy soils, where crop yields may be reduced by up to 50 per cent. Improperly cultivated peat soils with decomposed, powdery surfaces also have low resistance to wind erosion.

There is a strong seasonality in deflation with peaks in early spring and in summer. Improper farming practices may lead to a powdering of the soil surface or compaction, and ultimately to deflation.

In Hungary the major factor of wind erosion is the low cohesion of a dry soil surface. The obvious preventive measure is to ensure a proper *vegetation cover*, which reduces turbulent air motion on the surface. Rye sowing, mulching or green manuring are most often applied (Stefanovits, 1977). Inorganic materials are also suitable for sealing the soil surface (e.g. clay, bentonite injection, resins or plastic foils; see Szabó, 1977).

In order to allow mechanization after the collectivization of Hungarian agriculture, large arable fields were formed. Today where wind velocities are high and droughts are frequent, small (maximum 25 ha) plots separated by shelterbelts are recommended. Shelterbelts of trees of rapid growth (e.g. poplars and acacia) are preferred. Shelterbelts were introduced in Hungary following the Soviet example. They are now considered necessary on soils with poor water retention and which are liable to drought.

CASE STUDY: EROSION IN A SMALL WATERSHED

The area selected for closer study, the watershed of the Örvényes Séd stream, lies on the northern subcatchment of Lake Balaton. Soil erosion studies have been performed here for four years in cooperation between the German Research Foundation and the Geographical Research Institute, Hungarian Academy of Sciences. Among the objectives is the estimation of soil loss for each section of the watershed, using the Universal Soil

Loss Equation. Our research was also funded by Earthwatch (USA) and by the Hungarian Research Fund (OTKA).

The watershed has an area of 24.4 km^2 with a highest elevation of 416 m above sea level and an outlet at 104 m altitude (max. relief 311 m). The main water-course, the Örvényes Séd is 8.1 km long.

Geologically the area is built up of (karstic) Triassic limestones, dolomites and marls and their regolith (calcareous fragments embedded in clay matrix) as well as a mantle of unconsolidated deposits (loess, slope loess). On loess and marl the soils are either rendzinas or medium to severely eroded brown forest soils. In low lying areas meadow soils are found. On hilltops the parent rock is often exposed.

METHODS

According to Wischmeier and Smith (1978), the USLE expresses the rate of soil loss (t/ha), i.e.

$$A = RKLSCP,$$

where A is the computed soil loss (t/ha), R is the rainfall and runoff factor, K is the soil erodibility factor, L is the factor of slope inclination, C is the cover and management factor and P is the support practice factor.

In our case R and P are constants, since the methodological example is computed for a small part of a 20 km^2 catchment, i.e. for ca. 2 km^2. K and C factors were computed on the basis of available maps.

The LS factor was calculated for morphounits (small subcatchments) delineated on the contour map of the area. L and S were derived from DEM constructed by the method of triangular irregular networks (TINs).

The factors of the USLE were stored as GIS data levels. PC ARC-INFO was used for data storage and manipulation. Data storage in a GIS file makes on the one hand easy and quick calculations possible, on the other hand, it enables the user to manipulate the data levels (e.g. by overlapping).

The method is based on the definition of *erotops* (Fig. 1). An erotop is a unit for the calculation of soil loss. It is defined as a unit with approximately the same slope direction and landuse type and without rills of dells, i.e., without directed water discharge. The delineation of erotops was performed by Richter (1994). This method is, therefore, only GIS aided because only the calculation of soil loss was carried out within the frame of the GIS.

If we compare soil loss calculated for the whole catchment with our measurement data at Örvényes (see Table 2) we arrive at the conclusion that only ca. 2% of the calculated soil loss actually leaves the catchment.

Current changes in land ownership, reprivatisation, may influence sediment yields. As large fields are allotted to several smallholders, major landscape ecological changes are expected: plot boundaries may develop into artificial terraces or hedge rows, which probably further reduce the final loss of sediment and nutrients from the watershed.

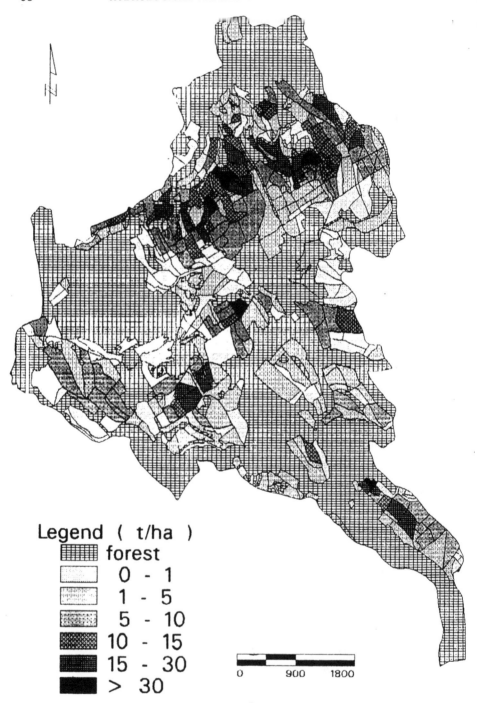

Legend (t/ha)

forest

0 - 1

1 - 5

5 - 10

10 - 15

15 - 30

> 30

0   900   1800

Fig. 1: Soil erosion map of Örvényesi Séd catchment.

Table 2: Sediment yield of Örvényesi Séd catchment (t/year) between 1977 and 1993 (calculations based on sediment measurements at Örvényes)

Year	Sediment (t)	Year	Sediment (t)
1977	68.05	1986	173.57
1978	45.97	1987	122.61
1979	58.97	1988	50.46
1980	18.49	1989	13.52
1981	73.79	1990	61.30
1982	80.16	1991	13.93
1983	177.64	1992	20.31
1984	864.2	1993	8.20
1985	91.45		

REFERENCES

Góczán, L. and Kertész, Á. 1988. Some results of soil erosion monitoring at a large-scale farming experimental station in Hungary, *Catena Suppl.,* 12: 175–84.

Hudson, N. 1971. *Soil Conservation*, Cornell University Press, Itchaca, New York.

Kerényi, A. 1985. Surface evolution and soil erosion as reflected by measured data. In: *Environmental and Dynamic Geomorphology*, Akadémiai Kiadó, Budapest, Studies in Geography in Hungary 17, pp. 79–84.

Kertész, Á. 1985. Subject and methodology of experimental geomorphology. In: Pécsi, M. (ed.), *Environmental and Dynamic Geomorphology*, Akadémiai Kiadó, Budapest, 21–9.

Péczely, Gy. 1979. *Éghajlattan (Climatology)*. Tankönyvkiadó, Budapest.

Pinczés, Z. 1982. Variations in runoff and erosion under various methods of protection. In: *Recent Developments in the Explanation and Prediction of Erosion and Sediment Yield. Proceedings of the Exeter Symposium*, July 1982. IAHS Publications 137: 49–57.

Stefanovits, P. 1977. *Talajvédelem-környezetvédelem (Soil conservation—environmental protection)*. Mezőgazdasági Kiadó, Budapest.

Stefanovits, P. and Várallyay, GY. 1992. State and Management of Soil Erosion in Hungary. Soil Erosion and Remediation Workshop, US—Central and Eastern European Agro-Enviromental Program, Budapest, April 27–May 1, 1992. Proceedings, RISSAC, Budapest, pp. 79–95.

Szabó, J. (ed.) 1977. *A Melioráció Kézönyve (Soil Amelioration Handbook)*. Mezőgazdasági Kiadó, Budapest.

Várallyay, G. 1986. *Soil Conservation Researches in Hungary,* Round Table Meeting on Soil Conservation Technologies 16–20, VI, USDA SCS-MÉM NAK, Budapest, pp. 5–8.

Várallyay, G. and Dezsény, Z. 1979. Hydrophysical studies for the characterization and prognosis of erosion processes in Hungary. In: *The Hydrology of Areas of Low Precipitation*, Proc. of the Canberra Symp., December 1979, IASH-AISH Publ. 128: 471–7.

Varga, J. 1985. *A temófold minóségvédelmének helyzete és a VII. ötéves tervi főbb feladatok (Soil conservation and the tasks in the 7th Five-Year Plan)*, Melioráció, Öntözés és Tápanyaggazdálkodás, 1, pp. 3–11.

Wischmeier, W.H. and Smith, D.D. 1978. *Predicting Rainfall Erosion Losses—A Guide to Conservation Planning:* USDA Agricultural Handbook 537. U.S. Government Printing Office, Washington, D.C., 58 pp.

Field Studies on Sediment Transport and Debris Flows in Small Basins of the Italian Alps

M.A. Lenzi, L. Marchi and P.R. Tecca

ABSTRACT

The paper summarizes recent research experiences conducted on sediment transport and debris flows in small streams of the Eastern Italian Alps. It illustrates the basic features of the gauging stations for sediment monitoring which are operating in two streams as well as the data collected. The results of field investigations on the volume of deposits from recent debris flows events are presented. Field studies on sediment transport and debris flows are discussed in the context of flow hazard mitigation.

Keywords: Sediment transport, debris flows, Italian Alps.

INTRODUCTION

In mountain basins of the Eastern Italian Alps (Fig. 1), different morphologic and landuse conditions occur. Hilly and low-mountain basins are found in the pre-alpine belt. These drainage basins are characterized by rather low elevation (seldom exceeding 1500 m a.s.l.); and their intermix of coppices and agricultural areas. Typical alpine conditions (very steep slopes, high-gradient streams, large portions of the basin area above the timber line) occur in mountain basins in the inner part of the range considered. Here, agricultural activity is usually limited to some favourable locations (wide valley bottoms, alluvial fans).

In the Italian Alps, depending on the variability of land use and the morphologic characteristics of drainage basins, different erosion and sediment transport processes take place, resulting in different problems in watershed management. The importance of sediment transport problems in the studied region is particularly obvious in the frequency of high intensity flow events that cause serious damage both in large rivers and in minor streams.

Most data on sediment yield in the region are obtained from reservoir surveys; in most cases these data concern quite large basins (with a drainage area usually exceeding 200–300 km²). Sediment monitoring in properly instrumented basins is needed to improve understanding of sediment transport processes in smaller catchments and, all the more, as a prerequisite for the implementation of sediment management programs. The occurrence of different sediment transport processes (water discharge

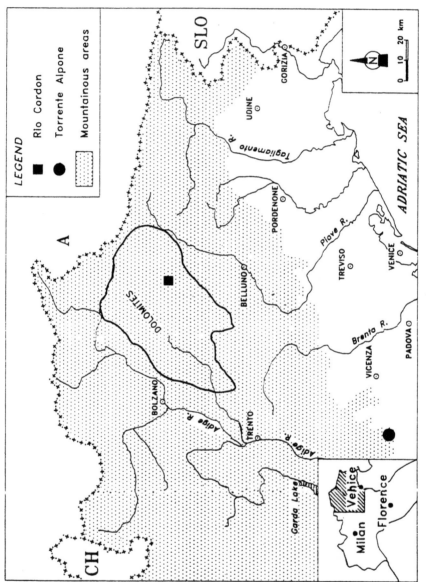

Fig. 1: Geographical location of the study area.

with prevailing suspended sediment, floods with massive bedload and debris flows) implies the need for different monitoring techniques in instrumented basins which are representative of physical conditions occurring in the Eastern Italian Alps. This paper briefly recounts some research experience.

SEDIMENT TRANSPORT MONITORING

A hilly basin representative of agro-forest catchments in the pre-alpine belt of the study area (the Torrente Alpone, basin area 77 km², average elevation 354 m a.s.l., range in elevation 872 m) has recently been instrumented. Precipitation, runoff and suspended sediment transport are being monitored (Cazorzi *et al.*, 1994). In this basin, whose location is shown in Figure 1, a gauging station was equipped such that the automatic sampling of flowing water occurs jointly with the continuous recording of suspended sediment concentration by surface scatter turbidimeter. The absence of direct contacts between samples and the optical parts lowers the drawbacks which are proper to many turbidimeters, the first of which is sensitivity loss by the instrument due to the deposition of sediment particles or organic matter on the sensitive parts.

Figure 2 shows water discharge and suspended sediment transport measured at the T. Alpone gauging station during a major flood event (Oct. 4, 1992). This event affected several parts of North-eastern Italy, causing widespread damage. In the T. Alpone basin, the return period of the precipitation that caused this flood event was estimated to be about 40 years; sediment yield far exceeds values recorded in other floods (Table 1).

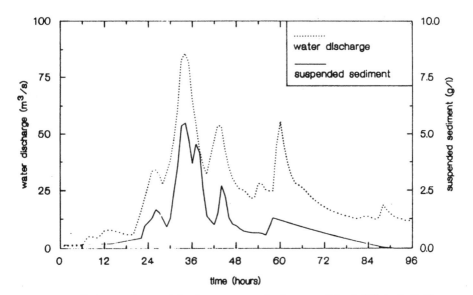

Fig. 2: Water discharge and suspended sediment transport for the event of Oct. 4, 1992 in the T. Alpone.

Table 1: Hydrological and sediment transport data recorded in the T. Alpone basin

Event	Storm rainfall (mm)	Measured runoff (mm)	Suspended sediment yield (t)
04/10/92	248.3	63.2	13083
03/12/92	40.5	1.8	47.5
05/12/92	46.6	1.8	714
08/12/92	104.7	26.6	2159
03/10/93	56.2	8.0	414
06/10/93	21.1	1.9	154
08/01/93	37.1	4.6	181

Researches currently carried on in the Alone basin concern the validation of hydrological and erosion basin models and the localization of specific areas within the basin affording high potential for soil loss and hydraulic risk. This was undertaken chiefly to determine the best management practices for controlling erosion and sediment yield.

Moving towards the inner part of the alpine range, the relative importance of bedload transport in minor streams usually increases, so that sediment transport measurement must take this component into account.

Traditional techniques for the measurement of total sediment transport in small mountain streams, such as sediment surveying in debris basins installed at the outlet of instrumented watersheds, show serious drawbacks and limitations. A novel experimental station for water discharge and sediment transport measurement has been designed (Fattorelli *et al.*, 1988) in order to obtain a continuous and more detailed measurement of different components of total sediment transport, including coarse bedload, in high-gradient alpine streams. This station was installed in the Rio Cordon (Fig. 1); the instrumented watershed drains an area of 5.0 km^2 (maximum basin relief 985 m, average elevation 2199 m). The absence of urban areas and roads in the Rio Cordon basin, where most basin area is assigned to pasturing and, to a lesser extent, to forestry, makes this basin suitable as a benchmark for comparison with other alpine basins affected by landuse changes.

Coarse sediment reaching the recording station installed on the Rio Cordon is separated from water and fine sediment by a grille provided with 20 mm distance between its bars, so that larger grade sediment slides over the grille and is deposited in an open storage area. The height of the deposit is measured here by a net of ultrasonic gauges; in this way it is possible to record the volume of coarse material deposited as the event progresses in time. Suspended sediment is measured by means of a light-absorption turbidimeter.

Figures 3 and 4 show water discharge and sediment transport measured during two flood events (D'Agostino *et al.*, 1993). This provides an example of data recorded at the Rio Cordon station.

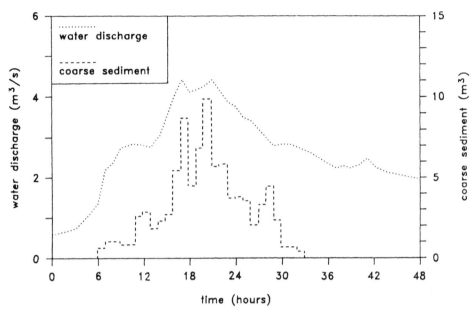

Fig. 3: Water discharge and coarse sediment accumulation during the event of July 3, 1989 in the Rio Cordon.

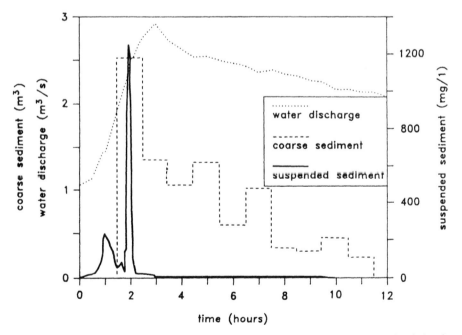

Fig. 4: Water discharge, coarse sediment accumulation and suspended sediment concentration during the event of October 5, 1992 in the Rio Cordon.

Researches currently carried on in the Rio Cordon basin concern refinement of sediment transport measurement, also by means of tracer techniques, the characterization of sediment source areas and hydrological basin modeling.

DEBRIS FLOW STUDIES

Floods with considerable sediment transport and debris flows commonly occur in small alpine watersheds. Debris flows, however, are recognised to be the more significant hazard to human life and works in these areas, since they are one of the most rapid, intense and destructive erosion processes. This form of flow process is intermediate between landsliding and waterflooding, with mechanical characteristics different from either of these processes (Costa, 1984).

While field observation systems for measuring the dynamic properties of debris flows have been developed for two decades in several geographical areas, as in Japan (e.g. Okuda et al., 1980; Suyama, 1988), limited experiences on experimental sites for debris flow studies have been conducted in the Alpine Region in recent years (Mortara et al., 1986; Blijemberg, 1993). However, many field investigations on debris flow in uninstrumented sites, reveal great interest in evaluating associated hazard, designing defence measures determination of main trigger mechanisms, delineation of transportation and deposition zones and estimation of the magnitude, i.e. the volume of debris delivered to the deposition area.

As to volume estimates of debris flows in the Eastern Italian Alps, catastrophic events, sometimes greater than 10^6 m^3, are documented in the literature (Baselli, 1923; Eisbacher and Clague, 1984). However, even minor-magnitude events have resulted in severe risk for urban areas and transit routes (Fig. 5): Table 2 shows deposited volumes and damage caused by some recent debris flow events in the studied region.

The evaluation of the distribution of debris flow deposits along slopes and channels provides a suitable complement to the estimation of debris flow volume in the terminal deposition area. Accumulated debris flow deposits can be estimated by means of topographic survey of lateral levees, debris deposited in flow diversions, medial deposits in channel and terminal deposits. Figure 6 shows deposits for some small-scale hillslope debris flows in the Dolomites. The relative importance of terminal deposits in total mobilized volume significantly varies from case to case. The variability in the pattern of accumulated debris flow deposits can be referred to differences in morphological conditions of debris flow tracks (in particular slope length and steepness).

CONCLUDING REMARKS

Field studies in instrumented basins and areas affected by recent flow events allow the collection of data on event frequency and magnitude and improve the knowledge of

Fig. 5: Interruption of a State Road (Rudavoi Creek, Dolomites, Aug. 1992).

Table 2: Debris flow magnitude and caused damage for some recent events in the Eastern Italian Alps

Event	Magnitude (m³)	Basin area (km²)	Damaged infrastructures
Rio Mulin, Sep. 1990 (*)	400	0.322	highway
Rio Badin, Sep. 1990 (*)	5000	0.306	highway
Campiolo, Sep. 1990 (*)	3700	0.070	highway
Rudavoi, Aug. 1992	5000	1.94	state road
State Road 51 K113.5, Oct. 1992	935	0.136	state road
Rio Bianco, Aug. 1986 (°)	6000–7000	6.64	highway state road
Fossa di Tovaccio, Aug. 1986 (°)	8000–10000	1.26	railway housing
Rio di Sacco, Aug. 1986 (°)	30000–40000	0.96	highway state road
Rio Dona, Jul. 1989	15000	2.86	urban area

(°) Mortara *et al.*, 1986
(*) Arattano *et al.*, 1991

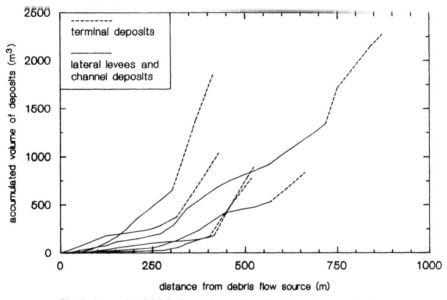

Fig. 6: Accumulated debris flow volumes along flow tracks in the Dolomites.

hazard conditions due to torrent activity. This is of the utmost importance for designing both structural and non-structural attenuation measures. As to non-structural measures, the delineation of hazard zones and the subsequent implementation of landuse restrictions appear to be the most suitable remedy in most cases. Structural measures for flow damage attenuation in minor streams (i.e. torrent control works) also benefit from reliable knowledge of the flow event expected (water floods with suspended sediment, heavy bedload transport, debris flows) and of the amount of sediment involved. Several offices of local authorities involved in mountain watershed management in the studied region have well established and long lasting experience in dealing with torrent hazard problems. However, the expansion of the main road network, the increasing importance of touristic activity in mountainous areas and associated landuse changes, the present threats of climatic changes demand a refined knowledge of flow processes in minor streams. Field researches, such as the ones reported in this note, can provide a significant contribution to this task.

ACKNOWLEDGEMENTS

The instrumented basins of the T. Alpone and of the Rio Cordon are managed by two offices of Veneto Regional authority (the Centre for Hydrology and Meteorology and the Centre for Avalanches and Soil Conservation respectively): the Directors of these Centres are kindly acknowledged for their cooperation. The investigations in the above-mentioned basins are carried on in the context of the EC programmes EPOCH

(T. Alpone) and the Environmental Research Programme—EROSLOPE project (Rio Cordon). Several people are involved in the researches in the instrumented basin: the authors wish to thank, in particular, Dr. G.R. Scussel (Veneto Regional Authority), Dr. M. Di Luzio and Dr. V. D'Agostino (University of Padova). Thanks also to Mr. G. Peruzzo (CNR IRPI—Padova) for drafting the figures.

REFERENCES

Arattano, M., Deganutti, A., Godone, F., Marchi, L. and Tropeano, D. 1991. L'evento di piena del 23–24 settembre 1990 nel bacino del Fella (Alpi Giulie). *Bollettino dell'Associazione Mineraria Subalpina*, 28 (4): 627–763.

Baselli, G. 1923. La catastrofe di Chiusa d'Isarco e le opere di riparazione e di prevenzione eseguite. *Giornale del Genio Civile*, 61: 619–634.

Blijemberg, H.M. 1993. Debris flow measurement methods. In: *Temporal occurrence and forecasting of landslides in the European Community. Final Report*, vol. 1, 147–159 pp.

Cazorzi, F., Diluzio, M. and Lenzi, M.A. 1994. Applicazione di un modello distribuito integrato con GIS per la simulazione dei processi idrologici ed erosivi nel bacino dell'Alpone. *Quaderni di Idronomia Montana*, 14 (in press).

Costa, J.E. 1984. *Developments and Applications of Geomorphology*. Edited by J.E. Costa, Fleisher P.J., Springer-Verlag, Berlin, 268–317 pp.

D'Agostino, V., Lenzi, M.A. and Marchi, L. 1993. Sediment transport and water discharge during high flows in an instrumented basin. In: *Dynamics and Geomorphology of Mountain Rivers* (K.H. Schmidt, P. Ergenzinger, eds.), Springer-Verlag, Berlin (in press).

Eisbacher, G.H. and Clague, J.J. 1984. Destructive mass movements in high mountains: hazard and management. *Geol. Survey of Canada, Paper*, 84–16, 230 pp.

Fattorelli, S., Keller, H.M., Lenzi, M.A. and Marchi, L. 1988. An experimental station for the automatic recording of water and sediment discharge in a small alpine watershed. *Hydrol. Sciences Journal*. 33, 6: 607–617.

Mortara, G., Sorzana, P.F. and Villi, V. 1986. L'evento alluvionale del 6 agosto 1985 nella vallata del fiume Isarco tra Fortezza ed il Passo del Brennero. *Memorie di Scienze Geologiche*, 38: 427–457.

Okuda, S., Suwa, H., Okunishi, K., Yokoyama, K. and Nakano, M. 1980. Observations on the motion of a debris flow and its geomorphological effects. *Z. Geomorph. Suppl.* 35: 142–163.

Suyama, M. 1988. Characteristics of debris flows and their breakup works in Japan. *Int. Symp. Interpraevent, Graz, 1988*, Vol. 2, 119–132 pp.

Deforestation of Amazonia: Global Climate Impacts

Luiz Carlos Baldicero Molion *

ABSTRACT

The Amazonas rainforest is being destroyed at high rates. Since the beginning of the 70s, the deforested area has increased by about 300 thousand km² which, if added to previous deforestation, amounts to approximately 415 thousand km², or 11.3% of the Brazilian Amazon Forest. The main causes for deforestation appear to be the government's geopolitical decision to occupy the region with the consequent opening of roads. The increasing population, as well as mechanisation of agriculture, in Southern Brazil are among the causes of high migratory fluxes into the Region. Large-scale deforestation may affect the global climate. The first hypothesis is that the Amazonas Forest plays an important role in the chemistry of the atmosphere, regulating the *Greenhouse Effect*; deforestation and burning may increase carbon-dioxide concentrations and thus enhance the *Greenhouse Effect*. The second hypothesis is that the forest is a source of heat for the atmosphere and regions outside of the tropics depend on the heat exported by Amazonia. Deforestation may reduce the power of the heat source, less heat would be available to be transported poleward, therefore affecting present climate stability. It is argued that these two hypotheses are not antagonistic but both contribute to global climate changes.

Keywords: Amazonia, deforestation, rain forest, climate change, greenhouse effect.

INTRODUCTION

Tropical forests are dynamical systems which have been disturbed naturally throughout the ages by climate fluctuations. At the end of last ice age, about 15,000 years ago, most of the humid tropics were colder and drier than at present and rainforests on several continents shrank to an area much smaller than their current range (Dickinson and Virji, 1987). On the other hand, during the so-called "*climatic optimal*", ca. 6,000 years ago, warmer and wetter climate conditions allowed the tropical forests to expand towards higher latitudes. The Amazonas Forest, for example, may have extended as far south as 25°S and eastward to the Atlantic coast. Tropical forests are also affected by short-term natural disturbances, such as interannual variability of climate and volcanic

*Permanent Affiliation: Instituto Nacional de Pesquisas Espaciais (INPE), C.P.515, 12.227-010, S.J. Campos, SP, Brazil.

eruptions. The Kalimantan Forest (Indonesian Borneo) had 3 million hectares destroyed by fires during the severe drought resulting from the 1982–83 Niño-Southern Oscillation event (Malingreau et al., 1985).

In recent years, mankind has been playing a major role in reducing the natural forest cover in the tropics through different forms of land use. The most serious destruction, it is said, is occurring in Amazonia, which is the largest expanse of tropical forest remaining in the planet. This paper reviews briefly the cases and the extent of Amazonian deforestation and focus the attention on its global climate impacts.

CAUSES, EXTENT AND RATES OF DEFORESTATION

The deforestation process in Brazilian Amazonia is highly complex; its reasons depend on the history of a given place and vary in different parts of the region. The primary driving forces seem to have been the geopolitical and economical decisions of occupying the Region and the increase of the regional population. The first has led to opening of highways—especially the Belém-Brasilia (BR-010) and the Cuiabá-Porto Velho (BR-364), along which the largest deforestation nuclei are found—and fiscal incentives policies, such as tax reductions and negative interest loans, which have benefited large enterprises and landholders. The regional population growth was mainly a consequence of the migration of small farmers, due to changes in agriculture patterns in southern parts of the country and to a lesser extent to real estate speculation. When compared to the above motives, mining and prospecting, subsistence agriculture and commerical logging are relatively minor contributors, although the latter may increase in the near future as the Asian and African forests are reduced. Detailed discussions on causes of deforestation are found, for example, in Fearnside (1983a; 1987), Hecht et al. (1988), Mahar (1979, 1988).

If there is a general consensus about the causes of deforestation, the same does not happen with respect to the total area deforested. The estimates are 600,000 km^2 (Mahar, 1988) and 660,000 km^2 (Myers, 1989), respectively 16.4% and 18.0% of the Brazilian Amazonas Forest taken equal to 3.66 million km^2 according to IBGE (1985) data. The disagreements emerge from the fact that these estimates are based on exponential projections of "desk surveyed" numbers. Brazil's National Institute for Space Research (INPE) published an inventory of alteration of forest cover evaluated from TM LANDSAT satellite imagery (Fearnside et al., 1990). The area altered by deforestation was about 415,000 km^2 or 11.3% of the Brazilian Amazon Forest.

In order to assess the impacts of deforestation on the global climate, it is necessary to know the present rates of destruction of tropical forests. Myers (1989) made a survey of deforestation rates which includes 34 countries that together comprise 97.3% of the existing tropical forests of the globe. According to this author's appraisal, during 1989 alone, tropical forests had lost 142,000 km^2 of their expanse. Together Brazil, Indonesia and Zaire account for 46 of this total. He estimated the current amount deforestation for the Amazonian countries (Bolivia, Brazil, Colombia, Ecuador, Peru and Venezuela) to be 66,000 km² per year and that Brazil was the overall leader, with a rate of 50,000 km² per year. As would be expected, rate assessments differ considerably from each other, and it may be possible they are overestimating the true deforestation. Myers' numbers for the Brazilian Amazon Forest alone, for example, are about 130% higher than INPE's average rate of 21,500 km² per year for the period 1978–89, but only 63% of the World Resources Institute (WRI, 1990) estimates, which were equal to 80,000 km². The projected deforestation rates, averaged for the period 1978–89, varied from 1.37 to 2.19%, while INPE's satellite appraisal was 0.59% per annum. The 1990–91 satellite survey resulted in a rate equal to 11,100 km² in that year (INPE, personal communication). This relative low rate is being attributed to the severe recession that Brazil is going through for the past 3 years.

GLOBAL IMPACTS

Due to its large extension, the Amazonas Forest may be important for the stability of the global climate. This section attempts to demonstrate this possibility, based on the present knowledge of both physical and chemical forest-atmosphere interactions. Two hypotheses have been developed: first, the forest as a *big filter*, controlling the *Greenhouse Effect*, and second, the region as an important *heat source* for the general circulation of the atmosphere.

(1) *Amazonia: the Big Filter*

In the carbon cycle modelling, there are large uncertainties regarding the role of the oceans and biosphere, particularly tropical forests, both in terms of storage and fluxes of carbon between these reservoirs and the atmosphere. Annually man's activities release into the atmosphere 5.4 ± 0.5 GtC (1 GtC = 10^9 or 1 billion metric tons of carbon) of fossil fuel burning plus 1.6 ± 1.0 GtC of deforestation, and subsequent burning. Of this total, 3.4 ± 0.2 GtC were estimated to remain in the atmosphere and 2.0 ± 0.8 GtC to be taken up by the oceans. The balance does not close, once 1.6 ± 1.4 GtC are *"missing"* (IPCC, 1990). Since the amount that remains in the atmosphere apparently is known with greater confidence, either the flux into the oceans is being largely underestimated (Tans *et al.*, 1990) or there is another important carbon sink which has been overlooked so far. The biosphere, both the temperate and tropical forests, is the best candidate. It has been observed at Mauna Loa that the amplitude of the CO_2 seasonal cycle has increased nearly 20% since 1958. This larger amplitude may be indicative of a growing net primary productivity of temperate forests or increased storage of carbon (Kolmaier *et al.*, 1989). The following argument raises the hypothesis that, presently, the tropical forests are also sequestering more carbon than previously and, thus, acting as an effective carbon sink during the entire year.

To study the influence that the Amazonian Forest exerts on the chemical composition of the atmosphere, two campaigns of the Global Tropospheric Experiment (GTE/ABLE) were made, one during the dry season, July–August 1985, (Harriss *et al.*,

1988) and another during the wet season, April–May 1987 (Harriss et al., 1990). Preliminary results indicated that the forest is a sink of tropospheric ozone (O_3) and the region as a whole is a source of methane (CH_4). It was also found that the forest had an average net daily carbon absorption rate, that is, photosynthesis minus plant and soil respiration, of 0.25 kilos of carbon per hectare per hour (Fan et al., 1990). Although these latter measurements were carefully made and analysed, they encompassed only 12 dry sunny days at the end of the wet season when, normally, trees are well supplied with soil moisture and can transpire freely. In these circumstances, one may argue that these measurements were biased towards optimal conditions and high carbon net intake rates which cannot be generalized for the annual cycle. The authors themselves are skeptical about their results. However, they also pointed out that, during the measurement period, the average quantum yield for absorbed PAR (photosynthetic active radiation) was only 30% of the theoretical maximum. In accordance, their numbers seem to be much lower than values quoted in the literature. Recently, Grace (1993) made similar measurements during 44 consecutive days in Jarú Forest, western Brazilian Amazon, and came up with the same number as Fan et al. (1990). Fung et al. (1987), in their Table 4, made a literature survey and assigned for tropical rainforest an average net primary productivity (NPP) of 945 gC m^{-2} yr^{-1}, which corresponds to 1.08 kgC ha^{-1} h^{-1}, that is, 4.3 times higher than Fan et al. (1990) and Grace (1993) reported values. In addition, in their annual cycle modelling, Fung et al. (1987) showed that the carbon uptake is virtually constant throughout the year over the Amazon Forest, with a secondary carbon uptake minimum exactly during the time of the Fan et al. (1990) and Grace (1993) field measurements. Thus, there is no reason to discard the 0.25 kgC ha^{-1} h^{-1} measured rate as a working average number. If it is done, this rate corresponds to a globally significant flux of carbon, about 1.2 GtC per year, over the 550 million hectares of forest. In other words, with present climatic conditions, the Amazonas Forest may be sequestering from the atmosphere the equivalent to 15% of the total carbon which is released annually through human activities.

　　Some scientists criticise this hypothesis under the argument that the forest is in a "climax" stage and, therefore, could not have a net absorption of carbon, because the gains through photosynthesis would be balanced by vegetation and soil respiration losses. This latter hypothesis is at least questionable, because it is known that photosynthetic activity depends on air temperature, availability of moisture and CO_2, variables which have not remained constant throughout the past decades. From late 19th century to the 1920s, the humid tropics were reported to be generally drier than the present. One of the causes was probably the higher frequency of the phenomenon El Niño-Southern Oscillation (ENSO), whose occurrence is known to be associated with below than normal rainfall in the tropics (Molion, 1990). From 1851 to 1930 occurred 9 strong and 16 moderate ENSO events, averaging one event per 3.2 years, whereas from 1931 to 1990, only 5 strong and 7 moderate events happened, that is, one event per 5 years (Quinn and Neal, 1987). In the period 1911–30, Manaus for example, in the middle of Amazonas Forest, had an average of 153.9 ± 19.7 rainy days per year; for 1961–90 period, the annual average increased to 189.3 ± 14.8 rainy days per year.

What is more important is that the dry season (Jul–Sep) also had a 34% increase of rainy days, being 19.6 ± 6.2 rainy days for 1911–30 period against 26.3 ± 7.0 rainy days for 1961–90, and became 32% wetter. The dryness of the beginning of the century was also apparent in the Negro River level, gauged at Manaus Port which reflects the behavior of that large forested basin. Between 1903–26, Negro River had 6 of its maximum yearly (flood) levels ranked amongst the lowest 10 maxima, the 1926 one being the *minimum maximorum* ever reported, over 5 standard deviations below the 90 year mean. In addition, the CO_2 concentration increased by 27% since that time. With higher water and CO_2 availability, tropical forests ought to grow more! In this respect, Goudrian and Unsworth (1990) suggested that an annual increase of 0.4% in the CO_2 concentration would stimulate an additional biomass storage of 1.5 to 2.5 GtC that, given the present uncertainties, are approximately the figures necessary to balance the carbon cycle. Recent results of a carbon balance model (Enting and Mansbridge, 1991) also suggested that tropical forests may be a larger carbon sink than temperate forests. It is possible, therefore, that tropical forests are increasing in biomass density, a parameter whose time variability is unknown. It could also be that forests are not retaining much of the carbon themselves but acting as a transfer mechanism from the atmosphere to other reservoirs, namely the soil and the aquatic systems, and the rivers, in turn, may be carrying part of the carbon to the ocean. The carbon in the ocean and in the soil has a higher residence time than that in the forest.

The destruction and burning of tropical forests, on the other hand, contributes to increasing the CO_2 and other greenhouse gases concentrations, enhancing the *Greenhouse Effect*. Currently, it is estimated that biomass burning in the tropics produces 1.6 ± 1.0 GtC annually, and that the Brazilian contribution is 540 MtC (WRI, 1990), that is, one-third of the total. These estimates may be much higher than the actual rates, both for Brazil and other tropical countries.

Releasing carbon by burning depends, initially, on the product of four parameters: the percentage of carbon in the biomass, the annual rate of deforestation, the forest biomass density and the efficiency of carbon release in the burning. The percentage of carbon in the biomass is taken equal to 45% as consensus. However, the average annual deforestation rates of the Brazilian Amazon, as it has been seen, differ widely among authors, from 2.1 million hectares (Fearnside *et al.*, 1990) to 8 million hectares (WRI, 1990). Anderson and Spencer (1991) found forest biomass density values ranging from 185 to 513 tons per hectare, while a literature survey indicated that the efficiency of carbon release by burning may vary from 20 to 100%. Multiplying these extreme values, one ends up with annual rates of carbon release that span from 35 MtC to 1.8 GtC, i.e., a factor of 52! Besides the percentage of carbon in the biomass, the Brazilian Amazon average annual deforestation rate is known more accurately today through satellite survey (Fearnside *et al.*, 1990). However, the uncertainty in the other two parameters, still gives a factor of 14 between the minimum and maximum estimated carbon release rates.

On the other hand, the contribution of Amazon biomass burning to the increase of CO_2 concentration may be relatively insignificant as far as the enhancement of the

Greenhouse Effect is concerned. According to Fearnside *et al.* (1990), the total deforested area in Amazonia is about 41 million hectares. Taking a biomass density of 300 tons per hectare (Brown, 1990) and supposing the worst conditions, that is, total release of the stored carbon, all carbon remaining in the atmosphere and no carbon fixation by regrowth, the transformation of Amazonas Forest into other land uses has contributed to increasing the atmospheric CO_2 concentration by about 5.5 GtC, i.e., 3.7% of the global contribution along the past 150 years, and 0.7% of the total carbon present in the atmosphere. In the remaining 550 million hectares of forest biomass, which includes all Amazonian countries, there is an estimated carbon storage of 74 GtC which, if all were released into the atmosphere, with no vegetation regrowth but part going into other sinks as considered today, would increase the CO_2 global concentration by a bit more than 5% over its present value, taken as 350 ppm. To put this in context, burning of fossil fuels releases carbon equivalent to a whole Amazonas Forest in less than 15 years!

In summary, the Amazonas Forest destruction hypothetically would contribute to the enhancement of the *Greenhouse Effect* in two ways: first, by biomass burning which is not too impressive when compared to fossil fuel burning by First World countries; second, by destroying the trees that transfer atmospheric carbon to the biomass and other reservoirs.

(2) *Amazonia: a Heat Source for the Atmosphere*

The solar energy reaching the surface is primarily used for evaporating water (latent heat) and heating the air (sensible heat). In Central Amazonia, micrometeorological studies (e.g. Molion, 1987; Shuttleworth, 1988) have shown that 80% ± 10% of the available energy are used for evapotranspiration (evaporation + plant transpiration), the rest warms the air. Over *terra-firme* forest, the water vapor flux is basically constituted of 70% transpiration of plants and 30% evaporation of rainfall water intercepted by the forest canopy and the litter layer. Direct soil evaporation was found to be negligible. In the annual mean, evaporation in Amazonia is about 50% of the total rainfall. In other words, assuming the climate is stable in the long term, half of Amazonia's rainfall comes from the local evaporation and the other half from the Atlantic Ocean (Molion, 1976; Salati, 1987). This local contribution is considered high when compared to what occurs in temperate latitudes, where it is estimated that local evapotranspiration makes up about 10% of local precipitation. When the water vapor condenses, forming clouds and rain, it releases the solar energy (latent heat) that was used in the evapotranspiration process.

Over a tropical continent, the warm and moist air rises (convection) and it is replaced, in the lower levels, by air coming from the oceans (convergence); in the high troposphere, the air is transported away (divergence) from the continent and sinks over regions near 30° latitude, thus closing a circulation cell. In the tropics, there are three regions where ascending motions predominate: the "*Maritime Continent*" (Indonesia, North of Australia), the Congo River Basin and the Amazonas River Basin, the latter two being of truly continental origin and, therefore, depending on the surface cover.

The first one is of different nature; it is a result of heat transfer from the ocean to the atmosphere as the Pacific waters in that region have surface temperatures of 29°C or higher. The latent heat released by these sources is transported away from the tropics to temperate and polar regions by the circulation cells, part of the general circulation of the atmosphere (GCA). Due to this transport, the global climate remains stable, but presents year to year variations which may be tied up to the fluctuating power of the sources.

The hypothesis is, thus, that the Amazonia is an important heat source for the GCA and large-scale deforestation may reduce the power of this source. As mentioned earlier, in the average, about 50% of the Amazonian rainfall comes from water evaporated locally. During GTE/ABLE-2 wet season campaign, Nobre *et al.* (1988) concluded that 58% of the rainfall, in fact, came from local evapotranspiration. Deforestation is known to reduce evapotranspiration, therefore, reducing precipitation and latent heat release. Experiments of large-scale deforestation of Amazonia were performed using global circulation models (e.g. Dickinson and Henderson-Sellers, 1988; Lean and Warrilow, 1989; Shukla *et al.*, 1990). The results showed a 20 to 30% reduction in rainfall (latent heat) over the basin, which corresponds to 9 to 13% of the total heat transported poleward by the atmosphere across latitudes 10° N and 10° S, based on data published in Hastenrath (1985).

Using a *climate feedback* equation proposed by Dickinson (1986), where the heat variation is equal to the temperature change times a *feedback parameter* (= 1.5 wm^{-20} C^{-1}), i.e., $Q = fT$, and assuming that only regions poleward of 30° latitude would be affected, such reductions in atmospheric heat transport to the extratropics correspond to reductions of 1.0 to 1.6 wm^{-2} or a mean temperature decrease of 0.7 to 1.1°C for these regions. One possible consequence of increasing the equator-to-pole temperature gradient is the equatorward displacement of the subtropical anticyclones (Flohn, 1981) and of the storm tracks (Paegle, 1987), conditions which are sufficient to change the present climate. Cooling also would shorten the growing season of countries located outside of the tropics, thus decreasing grain production.

Although the two hypotheses presented above, and the related arguments, are physically sound, the deforestation effects on the global scale remain controversial issues. One of the reasons is that the results of Amazon deforestation simulated with numerical models did not show impacts on the atmospheric circulation outside of the disturbed area.

SUMMARY AND CONCLUSION

The main causes of deforestation appear to have been the geopolitical decisions for occupying Amazonia which have led to the opening of roads and growth of population, mainly through migration to the region. Secondary causes, such as real estate speculation and fiscal incentives, contributed to the current deforestation rates. The principal land use after the removal of the forest has been pastures for livestock raising, which is the worst possible type of use for that environment.

The evaluation of landuse transformation in the Brazilian Amazonia, using LANDSAT TM images, suggests a deforested area of 415,000 km², which represents 11.3% of the area covered with forests, and a deforestation rate of 21,500 km² per year during the period 1978–89. Amazonia's deforestation has slowed down and, during 1990–91, rate was estimated to be 11.100 km² (INPE, personal communication).

The main objective of this work was to debate the two hypotheses of global climate change related to removal of the Amazonas Forest, namely the forest as a control of the *Greenhouse Effect* and the forest as a main source of heat for the extratropics. In theory, by burning Amazonia, the *Greenhouse Effect* might be accelerated. On the other hand, deforestation might reduce the power of the heat source, resulting in cooling of the extratropics and an equatorward displacement of subtropical anticyclones and storm tracks. If these hypotheses appear to be antagonistic, this just reflects the weakness of the present knowledge and the lack of appropriate tools to study the influence of that large forest on climate. The two unresolved questions should be addressed immediately, if global climate changes were to be understood.

Other consequences of deforestation should at least be mentioned briefly. Deforestation causes local climate changes. Soil surface temperatures may increase by 2 to 5°C and air temperatures by 1 to 3°C; evapotranspiration may decrease by 20 to 50% and local rainfall may be reduced by 20 to 30%. Another important component of the hydrologic cycle, the surface runoff, may decrease by 10 to 20% as rainfall is reduced (Dickinson and Henderson-Sellers, 1988; Lean and Warrilow, 1989; Shukla *et al.*, 1990). The GCMs results also suggest that deforestation may change both the temporal and spatial distribution of hydrologic variables and increase the length of the dry season, thus influencing the biota. Another problem, linked to the variation of the climatic elements and the removal of forests, is the soil degradation and consequent erosion due to the exposure of fragile soils to high rainfall rates. Jansson (1982) reviewed the literature on tropical soil erosion and found rates up to 334 metric tons per hectare per year. Erosion silts the river channels, lakes and reservoirs, changing the water quality and the aquatic life.

The great *biodiversity* of the region could be lost completely in a few years if present deforestation rates persist. It is said (see e.g., Mori and Prance, 1987) that the rainforest may contain 30 to 50% of the species of vegetation, animals and insects existing in the world. Although the forest is exuberant, 90% of the soil it stands on is said to be of old formation, leached and poor in nutrients. Closed nutrient cycle plays important role in maintaining the rainforest. Deforestation exposes the soil and the top 10–20 cm layer of organic matter is removed by weathering very quickly; soils cannot sustain agriculture or pasture for more than two to three years and, in the fallow land, the forest takes a long time to recover (Uhl *et al.*,1989). Changing the water quality and aquatic life by increasing sediment load and mercury concentration due to gold prospecting and mining, as well as increasing fertilizers and pesticides concentrations tied up with the expansion of agriculture frontiers, are among the problems that threaten the aquatic systems which contain 20% of the unfrozen fresh water of the planet.

If a conservative extent of population growth rate of 2% per annum is considered, one finds that, by the middle of next century, the world may have 12 billion inhabitants. It is obvious that a tropical region such as Amazonia, where in principle there are no climatic limitations for food production—even considering future scenarios predicted, for example, in case of enhanced *Greenhouse Effect*—should not remain marginal to this process. The development of the region, however, should be rational and careful in view of the arguments presented here. The question that follows is: "What is the most appropriate landuse for Amazonia?" Several authors (e.g., Goodland, 1980; Fearnside, 1983b; Uhl *et al.*, 1989) have dealt with this question. Molion (1986) also proposed some solutions for landuse in Amazonia. In conclusion, he wrote that the best solution seems to be a balance of natural forests, agricultural and pasture fields, being a higher proportion of forests and smaller of pastures.

REFERENCES

Anderson, J.M. and Spencer, T. 1991. 'Carbon, Nutrients and Water Balances of Tropical Rainforest Ecosystems Subject to Disturbances'. *MAB Digest* 7, 95 pp., UNESCO, Paris.

Brown, F. 1990. Personal communication.

Dickinson, R.E. 1986. 'Impact of human activities on climate—a framework'. In: *Sustainable Development for the Biosphere,* Clark, W.C. and Munn, R.E. (eds.) pp. 252–291, IIASA, Cambridge University Press.

Dickinson, R.E. and Virji, H. 1987. 'Climate changes in the humid tropics, especially Amazonia, over the last twenty thousand years'. In: *The Geophysiology of Amazonia*, Dickinson, R.E. (ed.), pp. 91–101, UNU, John Wiley and Sons.

Dickinson, R.E. and Henderson-Sellers, A. 1988. 'Modelling tropical deforestation: a study of GCM land-surface parameterization. *Quart. J. Roy. Met. Soc,* 114: 439–462.

Enting, I.G. and Mansbridge, J.V. 1991. 'Latitudinal distribution of sources and sinks of CO_2: result of an inversion study'. *Tellus* 43B: 156–170.

Fan, S.M., Wofsy, S.C., Bakwin, P.S., and Jacob, D.J. 1990. 'Atmosphere-biosphere exchange of CO_2 and O_3 in the Central Amazon Forest'. *J. Geophys. Res.* 95 (D10): 16,851–16,864.

Fearnside, P.M. 1983a. 'Landuse trends in the Brazilian Amazon Region as a factor in accelerating deforestation'. *Environmental Conservation* 10 (2): 141–148.

Fearnside, P.M. 1983b. 'Development alternatives in the Brazilian Amazon: an ecological evaluation'. *Interciencia* 8 (2): 65–78.

Fearnside, P.M. 1987. 'Causes of deforestation in the Brazilian Amazon'. In: *The Geophysiology of Amazonia.* Dickinson, R.E. (ed.), pp. 37–53, UNU, John Wiley and Sons.

Fearnside, P.M., Tardin, A.T. and Meira Filho, L.G. 1990. *Deforestation Rates in Brazilian Amazonia.* Instituto Nacional de Pesquisas Espaciais (INPE), S. José dos Campos, S. Paulo, Brazil.

Flohn, H. 1981. 'Scenarios of cold and warm periods of the past'. In: *Climatic Variations and Variability: Facts and Theories*, pp. 689–698, D. Reidel Pub. Co.

Fung, I.Y., Tucker, C.J. and Prentice, K.C. 1987. 'Application of Advanced Very High Resolution Radiometer Vegetation Index to study atmosphere-biosphere exchange of CO_2'. *J. Geophys. Res.* 92 (D3): 2999–3015.

Grace, J. 1993. 'Carbon cycle in Amazonia'. *Workshop on the Biosphere-Atmosphere Field Experiment in Amazonia.* INPE, São José dos Campos, S. Paulo, Brazil, 8–11 September.

Goodland, R. 1980. 'Environmental ranking of Amazonian development projects in Brazil'. *Environment Conservation,* 7 (1).

Goudriann, J. and Unsworth, M.H. 1990. 'Implications of increasing carbon dioxide and climate change for agricultural productivity and water resources'. In: *ASA Special Publication* 53, Madison, WI, USA.

Harriss, R.C., Wofsy, S.C., Garstang, M., Browell, E.V., Molion, L.C.B., McNeal, R.J., Hoell, J.M., Bendura, R.J., Beck, S.M., Navarro, R.J., Riley, J.T., and Snell, R.C. 1988. 'The Amazonas Boundary Layer Experiment (ABLE-2A): Dry Season, 1985', *J. Geophys. Res.* 93 (D2): 1351–1375.

Harriss, R.C., Wofsy, S.C., Garstang, M., Browell, E.V., Molion, L.C.B., McNeal, R.J., Hoell, J.M., Bendura, R.J., Coelho, J.R., Navarro, R.L., Riley, J.T., and Snell, R.L. 1990. 'The Amazonas Boundary Layer Experiment (ABLE-2B): Wet Season, 1987'. *J. Geophys. Res.* 95(D10): 16721–16736.

Hastenrath, S. 1985. *'Climate and Circulation of the Tropics'*, pp. 455, D. Riedel Pub. Co.

Hecht, S.B., Norgaard, R.B., and Possio, G. 1988. 'The economics of cattle ranching in eastern Amazonia. *Interciencia*, 13(5): 233–240.

IBGE. 1985. *'Anuário do Instituto Brasileiro de Geografia e Estatistica'*. Departamento de Estudos Geográficos, Brasilia, DF.

IPCC. 1990. *'Scientific Assessment of Climate Change'*. Intergovernmental Panel on Climate Change, Report of Working Group 1 chaired by Houghton, J.T., Seck, M., and Moura, A.D., WMO/UNEP, Geneva, Switzerland.

Jansson, M.B. 1982. *'Land Erosion by Water in Different Climates'*. UNGI Report n 57, Department of Physical Geography, Uppsala University, Sweden.

Kolmaier, G.H., Sire, E.O., Janecek, A., Keeling, C.D., Piper, S.C., and Revelle, R. 1989. 'Modelling the seasonal contribution of a CO_2 fertilization effect on the terrestrial vegetation to the amplitude increase in atmospheric CO_2 at Mauna Loa Observatory'. *Tellus*, 41B, 487–510.

Lean, J. and Warrilow, A. 1989. 'Simulation of the regional climatic impact of Amazon deforestation'. *Nature* 342: 411–413.

Mahar, D.J. 1979. *'Frontier Development Policy in Brazil: A Study of Amazonia'*, pp. 182, Praeger, New York.

Mahar, D.J. 1988. *'Government Policies and Deforestation in Brazil's Amazon Region'*. Environment Department Working Paper n. 7, pp. 42, The World Bank, Washington, DC.

Malingreau, J.P., Stephens, G. and Fellows, L. 1985. 'Remote sensing of forest fires: Kalimantan and North Borneo in 1982–83', *Ambio* 14 (6): 314–321.

Molion, L.C.B. 1976. *'A Climatonomic Study of the Energy and Moisture Fluxes of Amazonas Basin with Consideration of Deforestation Effects'*. INPE 923-TPT/035, pp. 119, São José dos Campos, S. Paulo, Brazil.

Molion, L.C.B. 1986. 'Landuse and agrosystem management in the humid tropics'. In: *Land Use and Agrosystem Management under Severe Climatic Conditions*. WMO Technical Note n. 184, pp. 114–137, WMO, Geneva, Switzerland.

Molion, L.C.B. 1987. 'Micrometeorology of an Amazonian rainforest'. In: *The Geophysiology of Amazonia*, Dickinson, R.E. (ed.), pp. 255–270, UNU, John Wiley and Sons.

Molion, L.C.B.: 1990. 'Climate variability and its effects on Amazonian hydrology'. *Interciencia* 15 (6): 367–372.

Mori, S.A. and Prance, G.T. 1987. 'Species diversity, phenology, plant-animal interactions, and their correlation with climate, as illustrated by the Brazil Nut family (Lecythidaceae)'. In: *The Geophysiology of Amazonia*, Dickinson, R.E. (ed.), pp. 69–89, UNU, John Wiley and Sons.

Myers, N. 1989. *'Deforestation Rates in Tropical Forests and their Climatic Implications'*. Friends of the Earth Report, London, U.K.

Nobre, C.A., Dias, P.L.S., Santos, M.A.R., Cohen, J., da Rocha, P.J., Guedes, R., Ferreira, R.N., and Santos, I.A. 1988. 'Mean large-scale meteorological aspects of ABLE-2B'. *EOS Transactions*, 69 (16).

Paegle, J. 1987. 'Interactions between convective and large-scale motions over Amazonia'. In: *The Geophysiology of Amazonia*, Dickinson, R.E. (ed.), pp. 347–390, UNU, John Wiley and Sons.

Quinn, W.H. and Neal, V.T. 1987. 'El Niño occurrences over the past four and a half centuries'. *J. Geophys. Res.* 92 (C13): 14,449–14,461.

Salati, E. 1987. 'The forest and the hydrological cycle'. In: *The Geophysiology of Amazonia*. Dickinson, R.E. (ed.), pp. 273–287, John Wiley and Sons.

Shukla, J., Nobre, C.A., and Sellers, P. 1990. 'Amazon deforestation and climate change'. *Science* 247: 1322–1325.

Shuttleworth, W.J. 1988. 'Evaporation from Amazonian rainforest'. *Proceedings R. Met. Soc. Lond.,* B 233: 321–346.

Uhl, C., Nepstad, D., Buschbacher, R., Clark, K., Kauffman, B., and Subler, S. 1989. 'Disturbance and regeneration in Amazonia: lessons for sustainable landuse'. *The Ecologist* 19 (6): 235–240.

Tans, P.P., Fung, I.Y., and Takahashi, T. 1990. 'Observational constraints on the global atmospheric CO_2 budget'. *Science* 247: 1431–1438.

WRI. 1990. '*World Resources 1990–91: A Guide to the Global Environment*'. UNDP/UNEP, Oxford Univ. Press, NY.

Effects of Acid Atmospheric Deposition on Watersheds in Central Europe

Josef Křeček

ABSTRACT

Acid atmospheric deposition is causing a great damage to the mountain watersheds in the Czech Republic. The contemporary forest decline in the Ore and Jizera Mountains is the result of synergism of a number of factors such as air toxicity, acidification of ecosystems and degradation of soil and low resistance of forest stands. Forest regeneration has been reduced by acidification of the habitats. The acidity of surface waters should be lower than that of precipitation, but in some cases it has exceeded that of rain. Control of these hazards is essential to protect such mountain watersheds.

Keywords: Air pollution, atmospheric deposition, watershed management, water quality.

INTRODUCTION

Acid atmospheric deposition is the most important factor affecting environmental changes in mountains of central Europe. The low buffer capacity of crystalline bedrock, poor shallow soils, and coniferous forests are typical for the mountain regions of the Bohemian Massif. Therefore, the values of the critical atmospheric load are relatively low (200–1000 eq ha^{-1} year^{-1}). They are being exceeded by 100–300% in the mountains on the border between the Czech Republic, Germany and Poland ("Black Triangle" of Europe).

AIR POLLUTION AND DEPOSITION

All the dominant sources of air pollution are connected with the use of fossil fuels. In the Czech Republic, annual burning of lignite increased from 10 to 100 megatons between 1945 and 1980, and lignite was the dominant source of energy production (78%) (the World Bank, 1991, Stevens, 1992). A similar situation has been observed in East Germany and Poland.

The highest deposition of sulphur (about 30 kg ha^{-1} year^{-1}) and inorganic nitrogen (about 20 kg ha^{-1} year^{-1}) in an open field was measured in the Ore Mountains in the eighties. Since 1988, a significant decrease in concentration of SO_2 in the atmosphere has been observed due to the international strategy of sulphur control, and political and economical changes in East Europe (Figs. 1, 2). The changes in atmospheric deposition in an open field at the Jizerka experimental watershed (the Jizera Mountains) are given

Air Control Stations
1—Kolová,	the Slavkovský Les Mts.,	575 m a.s.l.
2—Přebuz,	the Ore Mts.,	886 m
3—Bedřichov,	the Jizera Mts.,	820 m
4—Jizerka,	the Jizera Mts.,	870 m

Fig. 1: "Protected Areas of Natural Water Accumulation" in the Czech Republic: forested mountain regions of the total area 8,267 km² (30% of all the woodland).

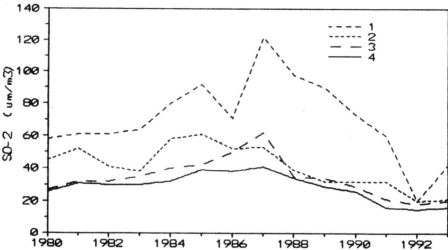

Fig. 2: Mean annual concentration of SO₂ in the air, observed at stations Kolová (1), Přebuz (2), Bedřichov (3) and Jizerka (4), in 1980–1993.

Table 1: Total annual deposition of major ions in the upper plain of the Jizera Mountains (Křeček, 1993)

Ions	Total annual deposition (tons km^{-2} year^{-1})			
	1986/87	1990/91	1991/92	1992/93
Sulphate	20.1	14.0	10.6	9.2
Nitrate	8.5	6.1	4.9	4.6
Chlorine	3.2	1.2	1.2	1.0
Fluorine	0.3	0.2	0.2	0.2
Ammonia	1.9	3.5	3.4	3.1
Calcium	3.1	1.2	1.3	1.2
Natrium	0.6	0.6	0.6	0.5
Magnesium	1.6	0.8	0.8	0.5
Sum	39.7	28.6	23.4	20.8

in Table 1. Under a forest canopy, the load increases with a canopy density and additional forms of precipitation (dew, fog, frost); an increase of 50–100%, in comparison with an open field, was observed in forest stands of Norway spruce (*Picea abies*) in the Jizera Mountains.

MOUNTAIN FORESTS

Significant human activities in mountain forests of the Czech territory started at the beginning of the 15th century. Priorities changed from game to timber harvest; by the 18th century. This resulted in the devastation of the forests in submountain areas caused by the commercial clear-cut, grazing of cattle, and harvest of litter or grass. The wildlife was systematically suppressed during the 18th century.

Subsequently, coniferous species (namely Norway spruce) have been preferred for the regeneration of forests (Table 2) for commercial reasons. This strategy led to a lower stability of mountain ecosystems. Already, at the end of the 19th century, several large insect epidemics occurred. Nowadays, 75% of mountain forests in the Czech Republic are significantly damaged by the consequences of air pollution, and 25% are devastated (World Bank, 1991). The contemporary forest decline in the Ore and Jizera Mountains (Table 3) is the result of a synergism of several factors (air toxicity, acidification of ecosystems and degradation of soil, lower resistance of forest stands, insect epidemics, and commercial forestry practices).

After logging, the mean annual erosion of forest soils increased from 0.1 to 1.4 mm in the Jizera Mountains, and the mean annual sediment output increased from 0.01 to 0.4 mm. These phenomena are consequences of soil erosion in skidroads that extended to 20–30 m ha^{-1} (Křeček, 1993).

Forest regeneration of large clear-cut areas is successful only in about 60%, being complicated by changes in bioclimate (snow and frost damages, water-logged soils

Table 2: History of tree species in the Jizera Mountains (Rabšteinek, 1969, in Křeček, 1993)

| Species | Percentage of tree species | | | | | | | |
| | Submountains | | | | Mountains | | | |
	0	1700	1810	1968	0	1700	1810	1968
Fir	42	30	13	0	23	33	17	0
Spruce	16	32	33	70	28	32	41	89
Pine	8	20	23	22	18	0	7	2
Beech	14	10	15	4	22	32	31	8

Table 3: Contemporary forest decline: clear-cut in watersheds of 4 drinking water reservoirs in upper plains of the Ore Mountains (Kamenička, elevation 596–885 m a.s.l.), and the Jizera Mountains (Bedřichov, 775–886 m, Josefodol, 733–1084 m, and Souš, 769–1122 m)

| Year | Area of forest clear-cut (%) | | | |
	Kamenička	Bedřichov	Josefodol	Souš
1965	8	1	1	1
1975	32	1	1	2
1985	65	8	3	26
1990	59	58	42	73

after the reduction of evapotranspiration), soil degradation, invasion of grass (mostly *Calamagrostis* sp. and *Avenella* sp.) (Fig. 5), and game damage (overpopulation of red deer).

WATER RESOURCES

"Acid" precipitation (pH lower than 5.0) has probably occurred for centuries, although acidity was not measured until 100 years ago. Renberg *et al.* (1993) identified four periods of pH development in natural lakes in southern Sweden during the postglacial time: 1) natural long-term acidification (until 2300–1000 B.C.), 2) anthropogenic alkalization (from 2300–1000 B.C. until 1900), 3) recent acidification (1900–1970), and 4) the liming period (since 1970s).

Reservoirs in the mountain regions of the Czech Republic have been constructed, since the end of the 19th century, to protect lowland cities against floods and, later on, for drinking water supply. The acidity of reservoirs in the Jizera Mountains was recognized already in the twenties, and attributed mainly to the dystrophic character of their watersheds (Stuchlík *et al.*, in press). In the sixties, a continuing decrease in pH, reduction in number of zooplankton species and disappearance of fish from the reservoirs were observed.

During the seventies and eighties, the pH of precipitation in the Ore Mountains as well as in the Jizera Mountains ranged between 3.2 and 4.0. Generally, the acidity of surface waters should be lower than that of precipitation, but in some cases the acidity of surface waters exceeded that of rain (throughfall and stemflow under the forest canopy, peat bogs and soils with a high content of humic acids). Therefore, the harvesting of forests (Table 3) led to an increase in pH values in streamflows.

On clear-cut areas in the upper plain of the Jizera Mountains, changes in the heat balance (an increase in albedo by 100% and in sensible heat flux by 40%, and a decrease in net radiation by 15%) led to a decrease in annual evapotranspiration by about 150 mm.

About 60% of the observed run-off resulted from fast sub-surface flow in podzolic layers of soil in the Jizera Mountains. After the clear-cut, the leaf area index (LAI) has declined from values of 16–20 to an annual course (described in Fig. 5), the infiltration capacity of soils dropped from 150 to 40 mm h^{-1}, and thus together with an extension of erosion rills after skidding of timber, changed run-off generation on hillslopes, increasing direct run-off by about 20%. These changes in run-off genesis could contribute to an epizodic acidification of reservoirs after heavy rainstorms or snowmelt events (Table 4, Fig. 4). Several of the parameters of the mountain reservoirs, low values of pH and water hardness, and high content of aluminium and humic acids, do not now meet the drinking water standard.

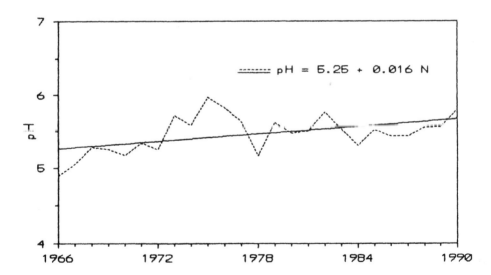

Fig. 3: Mean annual pH of run-off in the inlet of the Kamenička reservoir (the Ore Mts.) in 1966–1990 (Křeček et al., 1992). N—number of years since 1966.

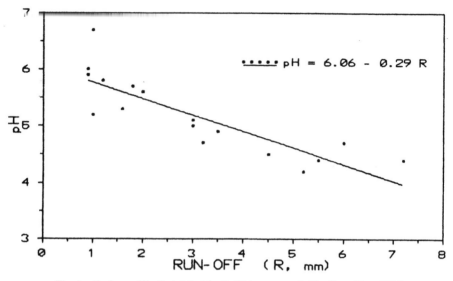

Fig. 4: pH of run-off in the inlet of the Bedřichov reservoir (the Jizera Mts., 1993)

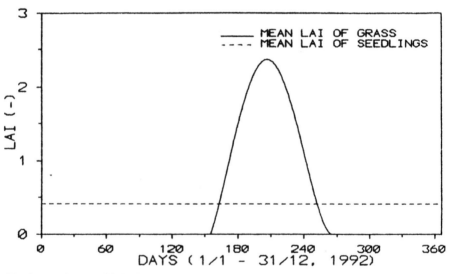

Fig. 5: Annual course of the leaf area index (LAI) in a clear-cut area, the Jizerka experimental basin (Křeček, 1993)

STRATEGY ON CONTROL OF THE MOUNTAIN ENVIRONMENT

The activities of nature conservation started to be realised in the 19th century. Two 'primaeval forests' (Zofin and Boubín) were established on the Czech territory in 1838 and 1848. After the World War II, an effort to protect individual trees, forest stands or

Table 4: Effect of storm run-off generation on water chemistry: direct run-off (surface and shallow subsurface water in an erosion rill) and base flow (deeper subsurface and ground water in a nearby stream channel in the Josefodol watershed (Křeček, 1993)

Ions	Concentration (mg^{-1})	
	Erosion rill	Stream channel
Sulphate	34.6	13.9
Nitrate	7.3	4.2
Calcium	8.8	4.5
Natrium	2.8	1.7
Magnesium	2.3	0.9

locations has been substituted by regional protection. According to the Nature Protection Act (40/1956), almost all the mountain regions in the Czech Republic have been proclaimed Protected Landscape Regions under special state care. A priority in water resources control in mountain regions (Fig. 1) was initiated by governmental acts in 1978 and 1979. These acts limited the clear-cut, deforestation, and drainage of forest soils, but were not respected in the practice. Unfortunately, the environmental aspects of timber harvest and regeneration technologies were not controlled at all.

International activities on air pollution control have led to a decrease in sulphur emissions in central Europe (Fig. 2). Although this reduction is very important, the critical load of acidity will be still exceeded in mountain regions of the Czech Republic for the near future. To find forestry practices adequate to maintain and rehabilitate mountain watersheds and lakes will be of special importance.

Nowadays, the political changes in the Czech Republic (more democracy and regional autonomy) seem to give a chance to control mountain forests on a regional principle, and to involve more interested groups (government, private forest owners, municipalities, local business activities, NGOs, and all the people living in the area).

ACKNOWLEDGEMENT

The investigation of mountain watersheds in the Ore Mountains and in the Jizera Mountains has been supported by the Earthwatch organization, Boston (U.S.A.) and Oxford (U.K.), within the project "Forests of Bohemia".

REFERENCES

Křeček, J. 1993. *Forests of Bohemia: rehabilitation of the air pollution damaged headwaters of North Bohemia.* Research Report, Earthwatch, 25 pp.

Křeček, J., Grip H., Navrátil, P., Skořepová, I., Tuček, M. 1992. Hydrology and biogeochemistry of forest decline effect in the Ore Mountains. In: Křeček, J. and M.J. Haigh (eds), *Proceedings of the Symposium on Environmental Regeneration in Headwaters,* WASWC/IUFRO, 93–99 pp.

Renberg, I., Korsman, T. and Anderson, N.J. 1993. A temporal perspective of lake acidification in Sweden. *Ambio* 22: 264–271.

Stevens, R.K. 1992. *Czech air toxicity study.* US-EPA/CZ-IHE, 12 pp.

Stuchlík, E., Hořická, Z., Prchalová, M., Křeček, J. and Barica, J., in press. Hydrobiological investigation of three acidified reservoirs in the Jizera Mountains, the Czech Republic, during the summer stratification. *Can. Tech. Rep. Fish. Aquat. Sci.*

The World Bank, 1991. *Czech and Slovak Federal Republic joint environmental study.* Report No. 9623-CZ, 131 pp.

The Effect of Clear Cutting, Waste Wood Collecting and Site Preparation on the Nitrate Nitrogen Leaching to Groundwater

Eero Kubin

ABSTRACT

Clear cutting causes considerable change in the nitrate content of groundwater. The content keeps on increasing for at least five years. Although the changes were big compared to the control area and to the situation before clear cutting, the amount of nitrate nitrogen remains relatively low when compared to the risk limit, above which drinking water is regarded as hazardous to the health. On the other hand, the leaching of nitrate out of the site may decrease the forest yield if it continues for a long time.

Plot studies show that, if the waste wood is collected, the change of the content is similar to the plot where waste wood was left. Site preparation, too, had a similar effect if there was no surface run-off. On an area where surface run-off appeared the content remained low for four years after ploughing.

Keywords: Boreal spruce forest, clear cutting, nitrogen, groundwater.

INTRODUCTION

Clear cutting causes considerable change in the structure of an unfelled forest and the interaction of its different growth factors. This becomes emphasized especially in the boreal spruce forests, which are shady and cool and, whose ground is covered by thick raw humus which contains a lot of nutrients (Kubin, 1983; Havas and Kubin, 1983). The clear cutting of such spruce forests increases considerably the temperature of the soil and the air layer near to it (Kubin and Kemppainen, 1991), which accelerates the decomposition of the humus, and the result is a drastic release of nutrients. A lot of nutrients are also released from the waste wood (Kubin, 1977). As the nutrient cycle between the tree stand and the soil stops completely after clear cutting, and there starts a new succession in the undergrowth, the nutrients which are released are for some time underutilized, and therefore prone to leaching. It has been shown that nutrients are leached after clear cutting both to surface and groundwater (Tamm *et al.*, 1974; Wiklander, 1974, 1983; Vitousek and Melillo, 1979; Rosén, 1983; Kubin, 1977; Ahtiainen, 1988).

In forestry, clear cutting is often followed by mechanical soil preparation. The result is that the temperature of the soil rises (e.g. Kubin and Kemppainen, 1992). The proportion of soluble nutrients in the humus inside the ploughing till grows, which leads to increased nutrient leaching in the surface water of the ploughed site (Kubin, 1987; Huttunen *et al.,* 1990). This is seen especially clearly in the increase of the nitrate nitrogen content, although the contents are considerably lower than what has been set as the risk limit for drinking water as hazardous to the health. The leaching of nitrogen can, however, have negative effects for the forest ecosystem, because the amount of mineralized nitrogen in the humus is usually small on the average or poor growing sites. In the northern areas, it has been found out that lack of nitrogen is a factor which limits the growth of tree stands.

This research examines the leaching of nitrogen to groundwater after clear cutting, waste wood collecting and site preparation. The leaching of nutrients to groundwater caused by forest treatment is, for the time being, rather inadequately documented.

EXPERIMENTAL METHODS

The experimental site is situated in northern Finland (Fig. 1). Two areas, whose tree stands differ a little from each other, were chosen as research objects. On the experimental site of Hautala the mean volume of the tree stand collected after clear cutting was 157 m³/ha, of which spruce formed 92%, and on the experimental site of Pahalouhi, 143 m³/ha and 47% correspondingly. The experimental sites also differ from each other in that on Pahalouhi there is surface run-off whereas on that of Hautala it does appear and especially on the ploughed experimental plot.

For the measurement of the height of the groundwater level and for examination of its chemical quality there were founded 49 groundwater wells in both areas. The well material was a plastic tube, the lower part of which was perforated for about the length of 1.5 m. The lower end of the tube was closed tightly with a plug. The measuring of the element contents of the water was started a year before the experimental sites were cut and prepared, which happened in the summer of 1986. The operations, the effects of which, on the quality of the groundwater, examined are: 1) clear cutting, 2) clear cutting and waste wood collecting, 3) clear cutting and soil preparation by ploughing, and 4) an unfelled control site (Fig. 1). The mechanical site preparation was done the same year as the clear cutting. All the clear cut areas were planted with pine seedlings in the following summer.

Water samples were collected monthly from June to October and the following element contents were measured: NO_3-N, NH_4-N, P-tot.K, Ca, Mg, Fe, Mn, Na and Zn as well as pH and acidity. In addition, the colour of the water was measured and the changes in its level were followed. In this paper, however, we examined only the effect of the operations on the nitrate nitrogen content and the pH value of the groundwater and their changes during five years after cutting.

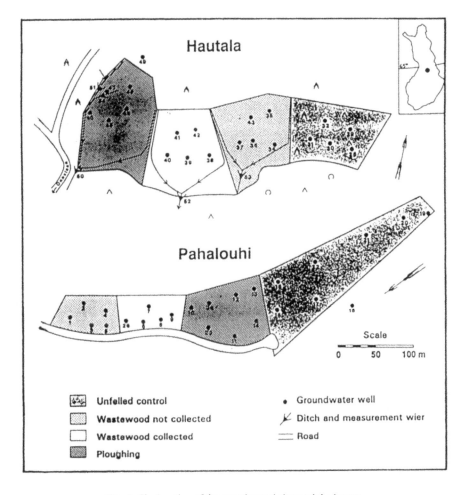

Fig. 1: The location of the experimental sites and the layout.

RESULTS

The effect of forest preparation was seen most clearly in the changes of the nitrate nitrogen content of the groundwater. Mere clear cutting, where the waste wood was not collected, nor was the soil prepared, caused a considerable growth of nitrate content in the groundwater by the second summer after cutting (Fig. 2) and the contents were higher than that in the three following summers. Between the successive years there was, however, only one statistical difference at the 5% level from the year 1988 to 1989. The effects were very similar on an adjacent experimental plot where the waste wood was collected immediately after cutting, but the comparison of data from year to year showed no statistical difference between successive years, although the mean values

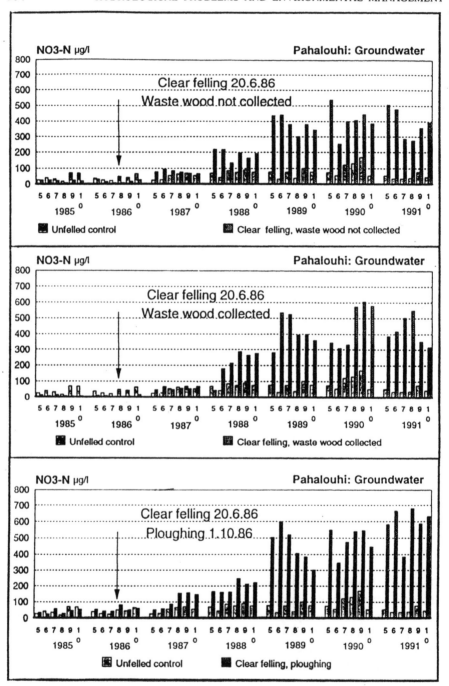

Fig. 2: The effect of clear cutting, waste wood collecting and ploughing on the nitrate content of the groundwater in the experimental site of Pahalouhi. Samples were collected monthly from June to October.

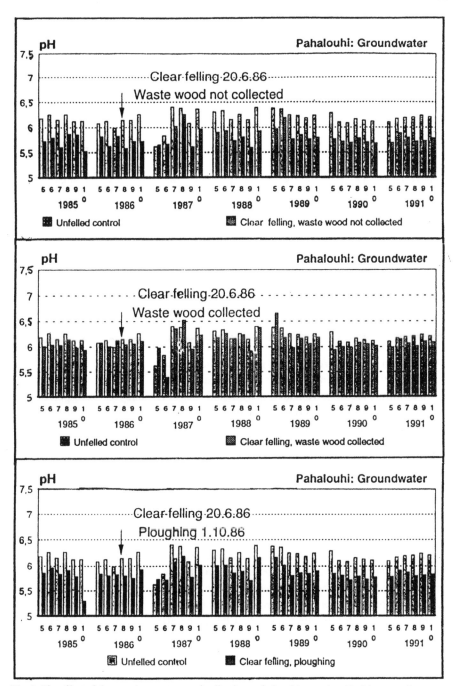

Fig. 3: The effect of clear cutting, waste wood collecting and ploughing on the pH of the groundwater on the experimental site of Pahalouhi. Legend see Fig. 2.

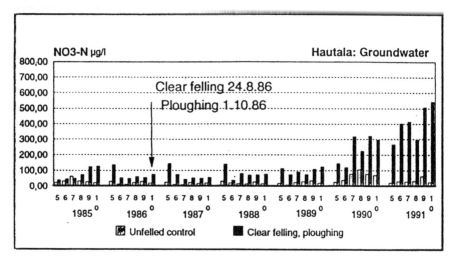

Fig. 4: The effect of ploughing on the nitrogen content of the groundwater in the experimental site of Hautala. Legend see Fig. 2.

showed quite clear differences. The case was similar on the ploughed plot according the mean values, but there the corresponding differences were observed between years 1988–89 and 1990–91. As for the pH (Fig. 3), there was observed only a slight increase in the acidity near the end of the period of measurement.

The nitrate leaching from the experimental sites of Hautala was corresponding on the unprepared plots. On the other hand, on the ploughed area, the contents in the groundwater remained low for three years after preparation and increased more considerably only in the third and fourth summers after preparation (Fig. 4). A statistical difference between successive years was observed only from the year 1990 to 1991.

CONCLUSIONS

In the experimental area, it has been shown that site preparation increases, among other things, the leaching of nitrate nitrogen in the run-off water in the spring (Kubin, 1987). On the other hand, neither a decreasing, nor an increasing trend was observed in the pH. Site preparation has been shown to cause nutrient leaching to surface waters in Nurmes research, too (Huttunen *et al.*, 1990). The effects of forest preparation on groundwater have not been examined in Finland before but in Sweden there are observations about nitrate nitrogen leaching to groundwater after cutting especially on sites with good quality (Wiklander, 1974, 1983). The results of this research are similar to the Swedish observations but the duration of the high contents seems to be longer and the changes of content smaller in this study. The quality of the experimental area of Kivesvaara is in the middle.

The finding that the content of nitrate grew in the groundwater both in the area where the waste wood was collected and where it was not, was generally not expected. It seems that the result is reliable because it was the same on both experimental sites. There are also unfelled and unfertilized forests above the experimental sites on both areas. More research is needed to find out the reasons for this similarity. The rise in the nitrate content caused by site preparation could be predicted because of earlier results, but the result, that on the area where there appeared surface run-off, the content did not increase until the fifth summer after ploughing, is interesting. The research on the duration of the changes of the content continues as well as that of other factors.

REFERENCES

Ahtiainen, M. 1988. Effects of clear-cutting and forestry drainage on water quality in the Nurmes-study. *Proceedings of The International Symposium on the Hydrology of Wetlands in Temperate and Cold Regions*. Joensuu, Finland 6–8 June, 1988. Vol. 1. Publications of the Academy of Finland 4/1988.

Havas, P. and Kubin, E. 1983. Structure, growth and organic matter content in the vegetation cover of an old spruce in northern Finland. *Annales Botanica Fennici* 20: 115–149.

Huttumen, P., Holopainen, A.L. and Ahtiainen, M. 1990. Avohakkuun ja maanmuokkauksen vaikutukset purojen veden laatuun ja vesibiologiaan. Joensuun yliopisto. Karjalan tutkmuslaitoksen julkaisuja 91.

Kubin, E. 1983. Nutrients in the soil, ground vegetation and tree layer in an old spruce forest in northern Finland. *Annales Botanica Fennici* 20: 361–390. Helsinki.

Kubin, E. 1987. Site preparation and leaching of nutrients. Proceedings of the IUFRO S1.05–12 Symposium Northern Forest Silviculture and Management in Lapland, Finland, August 16–22, 1987.

Kubin, E. and Kemppainen, L. 1991. Effect of clearcutting of boreal spruce forest on air and soil temperature conditions. Summary: Avohakkuun vaikutus Kuusimetsän lämpöoloihin. *Acta Forestalia Fennica* 225: 1–42.

Kubin, E. and Kemppainen, L. 1992. Influence of site preparation on air and soil temperatures and height growth of seedlings. Manuscript.

Rosén, K. 1983. *Mindre utlakning efter ristäkt*. Information från Projekt Skogsenergi. Nr 1: 16–17.

Tamm, C.O., Holmen, H., Popovic, B. and Wiklander, G. 1974. Leaching of plant nutrients from soils as a consequence of forestry operations. *Ambio* 3(6): 211–221.

Wiklander, G. 1974. Hyggesupptagningens inverkan på växtnäringsinnehåll i yt- och groundvatten. *Summary*: Effect of clear felling on the content of nutrients in surface and ground water. *Sveriges Skogsvårdsförbundets Tidskrift* 1: 86–90.

Wiklander, G. 1983. Kväveutlakning från Bördig skogsmark i södra Sverige. *Summary* : Loss of nitrogen by leaching from fertile forest soils in southern Sweden. *Skogs- och Lantbruksakademins Tidskrift* 122: 311–317.

Vitousek, P. and Melillo, J. 1979. Nitrate losses from disturbed forests: Patterns and mechanisms. *Forest Science* 25(4): 605–619.

Modelling the Hydrological Controls on Aluminium Leaching in a Forest Soil Impacted by Acid Deposition in Upland Wales

Chris Soulsby

ABSTRACT

Enhanced acid deposition on to the canopies of conifer forests in upland Wales has contributed to the release of Al into stream waters. Relatively little is known about how soil hydrological and chemical processes interact, in response to acid deposition, to transfer of Al-rich soil waters into streams. A hydrochemical field study monitored the hydrology and soil water chemistry of an afforested stagnopodzol at Llyn Brianne (Mid-Wales). Water moved vertically through the soil aided by root macropores. Simulated soil water fluxes indicated that 89% of effective precipitation drained vertically through the soil profile. Precipitation was acidified as water moved through the forest canopy and into the soil and Al was mobilized by cation exchange processes. Leaching of Al occurred during hydrological events and the estimated total flux was 3.39 kmol$_c$ ha^{-1} a^{-1}. It is likely that Al-rich soil water draining from the base of stagnopodzols provides a significant input into streams during acid episodes.

Keywords: Acidification, aluminium, hydrological pathways, afforestation, Wales.

INTRODUCTION

Deposition of acidic pollutants has been implicated in the acidification of soils and surface waters in afforested catchments in upland Wales (U.K.A.W.R.G., 1989). Consequently, forest streams can be acidic and Al-bearing with degraded aquatic ecosystems. The release of Al into streams is controlled by two principal factors: the soil chemical interactions governing the mobilization of Al, and the hydrological pathways which transfer Al-rich soil water into streams. Soil chemical processes have been examined in recent acidification research, but the role of hydrological pathways is often less well understood (Neal *et al.,* 1989). This paper presents data from an investigation carried out at Llyn Brianne in upland Wales, U.K. (Fig. 1). Field data was used to model soil hydrogeochemical processes and thus estimate water and solute fluxes from a forested stagnopodzol.

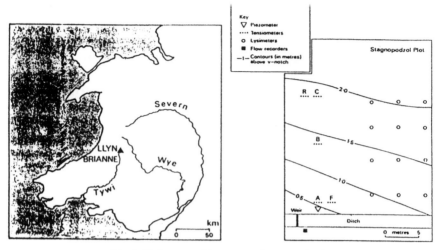

Fig. 1: Study area and instrumentation on study plot.

EXPERIMENTAL METHODS

A major catchment study was established at Llyn Brianne in 1984 to investigate the effects of acid deposition, afforestation and land management on stream acidification. Rainfall in the area has a mean pH of 4.9 and wet deposition of H^+ is 0.8 kg ha^{-1} a^{-1}. Extensive afforestation with Sitka spruce (*Picea sitchensis*) during the 1960s has increased the levels of acid deposition and during storm episodes forest streams have a low pH(<4.5) and elevated levels (>25 μmol dm^{-3}) of toxic inorganic Al species such as Al^{3+} (Goenega and Williams, 1988).

Stagnopodzols are an important soil type in upland Wales; they cover 20% of the afforested land at Llyn Brianne, and have been implicated as a probable source of acid, Al-bearing storm runoff. The soils have developed in periglacial drifts derived from underlying Lower Palaeozoic shale and are characterised by a 30 cm deep peaty O horizon above a 15 cm clay-rich Eag horizon. A thin ironpan separates these from the coarsely textured Bs and C horizons which merge into drift at 80 cm. The bedrock is within 1 m of the soil surface. An L horizon of spruce needles covers the well-structured O horizon where tree roots are abundant. Coarse tap roots penetrate the E horizon and the ironpan. The soils have been ploughed parallel to the slope prior to afforestation and drainage ditches dug at regular intervals.

A 5° hillslope plot was instrumented to study the hydrochemistry of an ironpan stagnopodzol (afforested in 1961) during the hydrological year 1988–89 (Fig. 1). Soil water pressure potentials were measured with a network of tensiometers which were monitored twice-weekly and hourly during storm events. Three sets of tensiometers were located in the lower (A), middle (B) and upper slope (C) of a hillslope transect. Replicate sets were placed in a plough furrow (F) and ridge (R). Tensiometers were located at depths of 25, 35, 55, and 75 cm to correspond to the O, E, B and C horizons respectively. Porous ceramic cups were used to sample soil waters from each horizon at fortnightly intervals. The cups were leached prior to installation and equilibrated *in situ* to avoid sample contamination (Hughes and Reynolds, 1990). Precipitation, throughfall and stemflow were measured close to the hillslope and fortnightly samples collected for analysis.

Samples were 0.45 μm membrane filtered. Inductively Coupled Plasma Emission Spectroscopy and Auto-analyzer techniques were used in the chemical analysis. Aluminium (measured by ICPES) was assumed to be monomeric (Al_m). This was fractionated into non-labile Al_m (mainly organically complexed; Al_o) and labile Al_m (mainly inorganic; Al) using the procedure described by Driscoll (1984). Unfiltered sub-samples were used for pH measurements. The physical (texture, bulk density, loss-on-ignition, saturated hydraulic conductivity (K(s)) and soil moisture characteristic curve) and chemical (pH and exchangeable bases, Al and H^+) characteristics of each soil horizon were determined using standard methods (Avery and Bascomb, 1974).

RESULTS

Soil Hydrology

The well-structured O horizon had the highest saturated hydraulic conductivity; permeability decreased in the E horizon and then increased in the coarse textured subsoil. All horizons were characterised by variability in K(s); this was associated with the location of macropores caused by root penetration (Table 1).

The soil water regime at each tensiometer set during the study year was similar to that of set B (Fig. 2). Rainfall totalled 1909 mm; interception losses accounted for 35 mm of precipitation, whilst 1391 mm and 183 mm reached the soil as throughfall and stemflow respectively. During the winter (days 1–182), pressure potentials remained above –6 kPa in each horizon. Detailed examination of storm hydrological pathways showed that the low slope angle and abundance of root macropores (penetrating the E horizon and ironpan) dictated that the profile was dominated by vertical drainage;

Table 1: Physical and chemical characteristics of stagnopodzol

Horizon	Mean K(s) (cm hr⁻¹)	pH	Exchangeable (μmol_c/kg)		
			H	Al	Bases
O	83.4 (8.3–167)	3.46	1.8	115	18.2
E	4.3 (2.3–8.0)	3.64	0.5	95.8	7.1
B	7.9 (0.9–20.8)	3.80	0.4	46.9	7.3
C	4.7 (1.9–19.0)	3.92	0.4	40.2	4.8

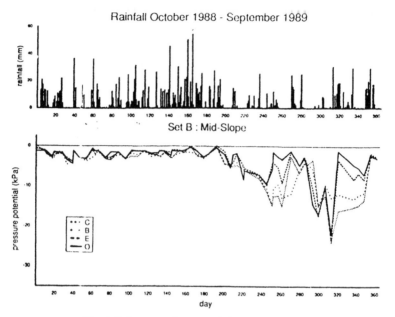

Fig. 2: Soil water regime at set B Sept 1988–Oct 1989.

lateral flow was only detected immediately above the shale bedrock (Soulsby, 1992). The marked drying of the soil during the summer commenced in May, with heavy rainfall in September returning the profile to pre-summer conditions.

Modelling Soil Water Fluxes

The tensiometer data facilitated the calibration of a simulation model that estimated soil water fluxes. The LEAching and CHemistry estimation Model (LEACHM) includes a soil water flow model that simulates vertical water movement by numerical solutions of the Richards Equation (Wagenet and Hutson, 1989);

$$\frac{\delta\Theta}{\delta t} = \frac{\delta}{\delta x}\left[K(\Theta)\cdot\delta h/\delta x\right] - U(x, t) \tag{1}$$

where Θ = water content (m³/m³), h = hydraulic head (mm), K (Θ) is the hydraulic conductivity (mm/d) at Θ, t = time (d), x is the depth (mm) and U is a sink term representing transpiration loss.

The model was parameterized for the upper 80 cm of the stagnopodzol profile using measured values of K(s), bulk density and soil retentivity; K(Θ) was calculated from these parameters (Hutson and Wagenet, 1991). The surface boundary fluxes were the daily rainfall totals and weekly estimates of transpiration loss. The model predicted soil water pressure potentials and water fluxes at 5 cm depth intervals for daily time steps. The model was calibrated using the first three months of tensiometer data using the mean pressure potentials in each horizon at sets A, B and C. The calibration was validated against data for the remainder of the year (Fig. 3). Measured pressure potentials initialized the model and values of K(s) were progressively reduced until simulated pressure potentials fitted those measured. The estimates of weekly water fluxes simulated for the soil profile were used to calculate an annual water budget which indicated that 89.3% of effective precipitation drained from the C horizon (Table 2).

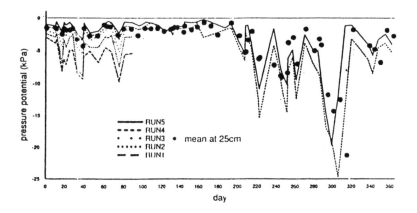

Fig. 3: LEACHM model calibration for stagnopodzol O horizon.

Table 2: Modelled soil hydrological budget for stagnopodzol 1 October 1988–30 September 1989

Precipitation input to soil	1575 mm
Transpiration losses from soil	196 mm
Drainage losses from C horizon	1407 mm
Change in profile water content	−28 mm

Soil Water Chemistry

The stagnopodzol was acidic with low base saturation and an exchange complex dominated by Al (Table 1). The chemistry of soil water reflected the enhanced deposition of mobile anions onto the forest canopy and the solid phase chemistry (Table 3); acidity was highest in the surface horizon with H^+ ions being consumed in the mineral subsoil. Aluminium was mobilized in O horizon and concentrations increased in the lower mineral horizons. Inorganic Al (Al_i) was the dominant fraction of Al_m with Al^{3+} the dominant inorganic species (Soulsby and Reynolds, 1992). The ionic composition of soil water was dominated by solutes of pollutant and marine

Table 3: Arithmetic mean concentrations ($\mu mol_c \, dm^{-1}$) of major solutes ($\pm$ SE) in soil solution

	O	E	B	C
pH	3.67	4.00	4.04	3.97
H	213 ± 12	101 ± 5	91 ± 6	107 ± 9
Al_i	91 ± 10	277 ± 15	317 ± 29	268 ± 20
Al_o*	8 ± 1	8 ± 2	9 ± 2	8 ± 3
Cl	498 ± 29	423 ± 16	518 ± 63	303 ± 10
SO_4	264 ± 22	262 ± 11	280 ± 21	243 ± 10
NO_3	9 ± 5	31 ± 5	34 ± 11	55 ± 7
Na	445 ± 35	320 ± 15	399 ± 41	248 ± 11
K	13 ± 6	19 ± 8	7 ± 2	5 ± 2
Ca	20 ± 3	31 ± 6	25 ± 4	19 ± 2
Mg	98 ± 6	72 ± 3	69 ± 4	36 ± 1

*($Al_o = \mu mol \, dm^{-1}$)

origin. Sulphate and Cl accounted for over 90% of the anions, with Na the dominant cation.

Aluminium Mobilization

Although Al is ultimately derived from the weathering of aluminosilicate minerals, it is extremely unlikely that mineral solubility controls regulate Al_i levels in soil waters in upland Wales (Neal *et al.*, 1989). It has therefore been suggested that cation exchange reactions may control the short-term mobilization of Al_i (Reynolds *et al.*, 1988). Cation exchange reactions can be simply described by a non-specific thermodynamic exchange equation involving a solid phase of constant exchange capacity;

$$Ads(M) + mN^{n+} = Ads(N) + nM^{m+} \tag{2}$$

where M and N are cations with respective charges m^+ and n^+, and Ads ( ) is the exchangeable fraction of each. Over a short timescale the number of exchange sites greatly exceeds the number of exchangeable cations passing through in the soil water. Thus, the ratio of adsorbed ions remains virtually constant so that ratios between cation pairs in solution also remain constant;

$$(M^{m+})^n/(N^{n+})^m = constant \tag{3}$$

Constants can be calculated for the ratios between Al^{3+} and other cation species and a theoretical relationship between the concentration of a given cation and the total anionic charge. The predicted concentrations of Al^{3+} and other cations agreed with those measured (Fig. 4), implying cation exchange relationships hold over the observed range of anion concentrations.

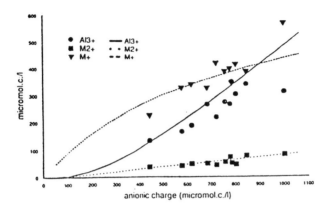

anionic charge (micromol.c./l)

Fig. 4: Predicted (lines) and actual (dots) cation concentrations in C horizon soil water.

Aluminium Leaching Fluxes from the Soil Profile

Fortnightly water fluxes estimated for each soil horizon were multiplied by solute concentrations to calculate the mass flux from the soil profile. In view of the uncertainties about the accuracy of the soil water fluxes, these must be regarded as preliminary (Table 4). The vertical flow paths result in leaching below the C horizon. The H^+ fluxes show losses from the O horizon and neutralization with depth. This was accounted for by increased Al_i fluxes. The flux of anions remained fairly constant and losses indicated the ability of mobile anions to leach Al_i from the profile. The excess cation flux reflects the loss of organic anions. The Al_i loss from the soil is just over 1% of the estimated pool of exchangeable Al. Moreover, calculated weathering rates show that the release of Al from minerals within upland soils on Lower Palaeozoic rocks is probably in excess of the estimated annual leaching losses (Evans and Adams, 1975). Thus, there is little evidence to suggest that acid deposition will cause significant depletion of the pool of exchangeable Al.

Table 4: Leaching fluxes ($kmol_c$ ha^{-1} a^{-1}) of hydrogen, aluminium, base cations (ΣBC), total cations (ΣC) and total anions (ΣA) and pool of exchangeable Al in stagnopodzol ($kmolcha^{-1}$)

	H	Al_i	ΣBC	$.\Sigma C$	ΣA	Al pool
O	2.70	0.95	6.56	10.2	9.27	112.3
E	1.28	3.15	5.27	9.70	9.70	98.3
B	1.06	3.36	5.37	9.79	9.19	52.5
C	1.27	3.39	3.66	8.32	7.70	42.4

CONCLUSIONS

These results demonstrate that Al_i leaching in acid soil profiles is strongly controlled by hydrological factors. Although cation exchange processes are probably the dominant chemical process governing Al_i, release, Al is transported through and from the soil along the dominant hydrological pathways. Below the soil profile, water is deflected downslope above the impermeable shale bedrock towards drainage channels. Buffering reactions in the drift below the soil are limited and drainage water probably provides a significant source of acid, Al-bearing storm runoff in afforested catchments at Llyn Brianne (Soulsby, 1992). Despite proposals by the European Community to reduce S emissions by 70%, the enhanced deposition of pollutants on to forest canopies in upland Wales will offset the effects of this directive and it is likely that SO_4 inputs will contribute to Al leaching from acid forest soil in the foreseeable future. Thus, acid stream waters are likely to remain a problem in upland Wales unless afforestation is restricted in acid-sensitive catchments.

REFERENCES

Avery, B.W. and Bascomb, C.L. 1974. *Soil Survey Laboratory Methods.* Soil Survey Technical Monograph No. 6, Harpenden.

Driscoll, C.T. 1984. A procedure for the fractionation of aqueous aluminium in dilute acidic waters. *Int. J. Anal. Chem.* 16: 267–283.

Evans, L.J. and Adams, W.A. 1975. Quantitative pedological studies on soils derived from Silurian mudstones 5. Redistribution and loss of mobilized constituents. *J. Soil Sci.* 26: 327–335.

Goenega, X. and Williams, D.J. 1988. Aluminium speciation in surface waters from a Welsh upland area. *Environ. Pollut.* 52: 131–150.

Hughes, S. and Reynolds, B. 1990. Evaluation of porous ceramic cups for monitoring soil-water aluminium in acid soils. *J. Soil Sci.* 41: 325–328.

Hutson, J.L. and Wagenet, R.J. 1991. Simulating nitrogen dynamics in soils using a deterministic model. *Soil Use Manage.* 41: 74–78.

Neal, C., Reynolds, B., Stevens, P.A. and Hornung, M. 1989. Hydrogeochemical controls for inorganic aluminium in acidic stream and soil waters at two upland catchments in Wales. *J. Hydrol.* 106: 155–175.

Reynolds, B., Neal, C., Hornung, M., Hughes, S., and Stevens, P.A. 1988. Impact of afforestation on the soil solution chemistry of stagnopodzols in mid-Wales. *Water, Air & Soil Pollut.* 38: 55–70.

Soulsby, C. and Reynolds, B. 1992. Modelling hydrological and Al leaching in an acid soil at Llyn Brianne, mid-Wales. *J. Hydrol.* (In Press)

Soulsby, C. 1992. Acid runoff generation in an afforested headwater catchment at Llyn Brianne, Mid-Wales, *J. Hydrol.* (In press)

United Kingdom Acid Waters Review Group (U.K.A.W.R.G.). 1989. *Acidity in United Kingdom Fresh Waters,* H.M.S.O., London.

Wagenet, R.J. and Hutson, J.L. 1989. *Leaching Estimation and Chemistry Model.* Water Resources Institute, Cornell University, Vol. 2, pp. 1–148.

Air Pollution Deposition Variability in a Slovene Alpine Headwater as a Consequence of Topography and General Circulation Assessed by Theoretical Model

T. Vrhovec, N. Pristov and A. Hočevar

ABSTRACT

Air flow over complex topography is irregular with eddies and local flows created or modified by ridges and valleys. Air pollution deposits are to a great extent dependent on it. Therefore, we built a model to gain a quantitative description of the wind field. With the help of a diagnostic mesoscale model (variational, with geostrophic and mass consistency constraints) we calculated three-dimensional wind field for an alpine headwater with its prescribed topography and upper air flow. Properties of the air, air pollution and others, are carried and deposited according to wind velocity distribution. With the help of a simple pollution transport, dissipation and deposition model ground level pollution distributions were calculated. Results of the simulation explain the very large variability of air pollution deposits in the waterhead studied. Models of this kind can be very helpful in the analysis of air pollution deposit distribution in an irregular terrain and at planning of its monitoring.

Keywords: Air pollution, topography effect, alpine headwater, Slovenia.

INTRODUCTION

Transport of air pollution at various spatial scales has been intensively studied for a long time (Anonymous, 1992; ApSimon, 1985, 1987; CCU, 1973, and many others). As one of the results of such studies including its longe range transport of pollutants (measured by surface wet and dry deposition) in Europe were calculated as a function of air pollution emissions (OECD, 1974, 1979; NILU, 1984) and general circulation of the atmosphere.

The potential importance of the relative vigour of westerly air flow in such latitudes in relation to topography can be demonstrated by examination of the long record of aerosols sulphate and nitrate measured at the rural site of Chilton in south-central England (UKRGAR, 1990).

Models have been built to predict gas plume behaviour as a function of different factors, for example, the source of the emissions, wind velocity, turbulence, atmospheric stability (Bosanquet *et al.*, 1950), stack height (Rakovec and Petkovšek, 1975) and buildings. In microscale studies of the deposition environment, velocities of air pollutants have been assessed using observed profiles of wind velocity and concentration of various pollutants and resistance analogs (Sutton *et al.*, 1992).

Studies regarding particular basins have been made in Slovenia, as well (Petkovšek, 1974; Hrček *et al.*, 1988; Završek, 1989, and others). An extensive study of long-range transport into and from Slovenia was made by Petkovšek (1974). Similar studies were made also for the whole alpine region and for Dinaric alps (Klaic, 1990).

In our present paper we develop a new point of view regarding the air pollution problem in complex topography. This concerns the way wind velocity distribution—which in irregular terrain is uneven influences air pollution deposits and gaseous air pollution concentrations in an idealised watershed. The resolution of this question may be of great help in the analysis of air pollution deposit distributions in complex terrain and for planning of air pollution monitoring.

WIND MODELLING IN A HEADWATER AREA

For determination of the wind field in a topographically complex terrain, a model is necessary, especially if we want to deal with realistic topography and real measured data. We used a diagnostic wind model (Rakovec *et al.*, 1992; Vrhovec, 1988, 1988a; Sherman, 1978).

The objective of this model is to reconstruct the wind field in complex terrain using the available data from a regular observational network. For upper air data in a scale between meso-γ and meso-β, as in our case it is typical to have only one or two soundings in the area (see Fig. 1) and from 10 to 20 wind data at ground level (see Fig. 2). The scale of the model grid is in the meso-gamma domain, so the typical resolution of the grid is about 3.5 km.

At upper levels, variational analysis with geostrophic constraint is applied (Andersson *et al.*, 1986).

The quality of the final results depends on the quality of the initial interpolation. First, it is necessary to determine approximately on which level it is expected to find the free atmosphere conditions for the wind. It is obvious that this level is higher in the area of mountains and lower over flatlands.

From the level of the geostrophic wind the wind is interpolated downwards using the Ekman profile, but not just to the ground; at some level above the smoothed terrain, the wind is allowed to turn towards the wind measured at ground stations. After vertical interpolation is made, incompressible (mass-consistent) variational analysis is used to correct the interpolated initial wind field (Rakovec *et al.*, 1992; Vrhovec, 1988; Sherman, 1978).

The model is constructed in a block coordinate system, so it can be used on topography with very steep slopes.

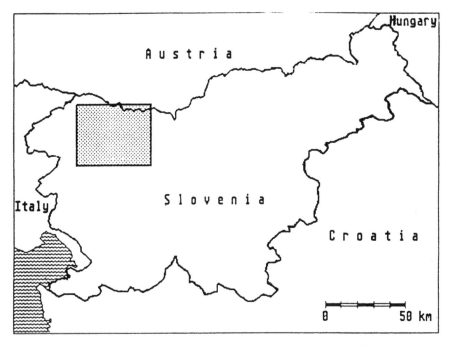

Fig. 1: The map of Slovenia and surrounding states; the model area is shaded.

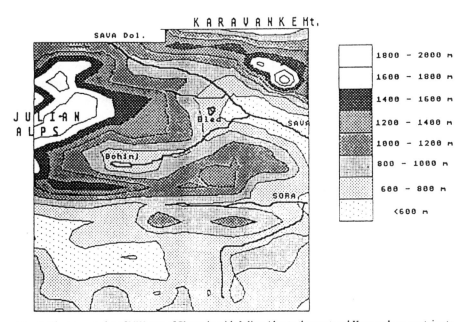

Fig. 2: The topography of NW part of Slovenia with Julian Alps to the west and Karavanke mountains to north with main rivers.

GEOSTROPHIC VARIATIONAL ADJUSTMENT ANALYSIS OF UPPER LEVEL WIND AND MASS-FIELD

Defining the initial values of the upper level wind, in a fine grid on mesoscale ($1\,km < \Delta x < 10\,km$) with few upper air soundings only, can be done by simple extrapolation. The initial fields are necessarily very smooth and have no small-scale structures. In our case the nearest upper air soundings are at Zagreb (Croatia) and Udine (Italy) and the fine grid values are determined by bilinear interpolation, using these and two additional Middle-European stations.

To achieve a consistency between the upper air wind field and the mass-field (geopotential), a variational approach similar to that of Andersson *et al.* (1986) was used.

Defining u_i, v_i and ϕ_i as first guess (initial, interpolated) fields and u_g, v_g and ϕ_g as adjusted fields a function is formed:

$$\iint_s [\alpha(\phi_g - \phi_i)^2 + (1 - \alpha)\,\beta(u_g - u_i)^2 + (1 - \alpha)\,\beta(v_g - v_i)^2]dxdy \qquad (1)$$

where α is a weight defined by mean square interpolation error of geopotential and β is a ratio between the typical values of geopotential and wind speed. The strong constraint (Sasaki, 1958) is introduced by geostrophic balance:

$$u_g = -\frac{1}{f}\frac{\partial \phi_g}{\partial y} \qquad\qquad v_g = -\frac{1}{f}\frac{\partial \phi_g}{\partial y} \qquad (2)$$

where f denotes the Coriolis parameter.

The function is minimised over the complete analysis area. The result of the minimisation procedure is an Euler-Lagrange equation. The lateral boundary conditions are determined from constraints while the geopotential is supposed to be exact and is not varied at the boundaries. With this lateral boundary condition, the equation can be solved either by a successive over-relaxation method or by a spectral method—both methods proved to be successful.

HORIZONTAL AND VERTICAL INTERPOLATIONS

10 to 20 measured values of the ground level wind can be expected to be routinely available in the area of interest. For horizontal interpolation of surface wind, a modified distance, which apart from cartesian distance also takes into account the variance of the terrain, is used (Vrhovec, 1988).

The use of such corrected distance enables us to reduce the strong influence of near points which, although geometrically each close to another, may be meteorologically far if they are separated by a mountain barrier. Additionally, different weight is given to the vertical and horizontal component of the distance. The vertical component of the distance is enhanced.

For vertical interpolation from the level of the geostrophic wind towards the ground, first a rough estimate is made. For this purpose, the so-called resistance laws (e.g. Tennekes, 1973; Mełgarejo and Deardorff; 1974; Zilitinkevich, 1975) can be applied in homogeneous terrain. For the sounding of Zagreb or Udine the homogeneity can approximately be justified. For the estimation of stability of the PBL (Planetary Boundary Layer) two layers are used: the 1000 mbar/850 mbar and the 850 mbar/700 mbar. The appropriate Richardson's number is used to determine the two functions A (μ) and B (μ), using semiempirical relations between these two functions and the Richardson number (Melgarejo and Deardorff, 1974; Zilitinkevich, 1975).

In neutral atmosphere, the height of the planetary boundary layer h_{PBL} can be determined iteratively from the two diagnostic equations:

$$\frac{U_g}{u_{*0}} = \frac{1}{k} \sqrt{\left(\ln \frac{h_{PBL}}{z_0} - B(\mu)\right)^2 + A^2(\mu)} \tag{3}$$

$$h_{PBL} = \text{const} \frac{u_{*0}}{f} \tag{4}$$

Here U_g is the velocity of geostrophic wind: $U_g = \sqrt{u_g^2 + v_g^2}$, k is the von Kármán constant, f is Coriolis parameter, u_{*0} friction velocity, and *const* some numerical constant, being very close to the von Kármán constant.

From the theoretical point of view the use of the second equation is not justified in non-neutral cases, but in this study it is still pragmatically applied (for the sake of simplicity and to avoid the introducing for example some PBL model for determining the height of PBL, like in Rakovec, 1983).

In complex topography, it is not appropriate to extrapolate the Ekman profile just to the ground. In valleys, the actual (measured) wind is often strongly channelled. The Ekman profile is matched at the height $h_s + \Delta h$ smoothly with a polynomial profile leading to the value of the wind at the ground. This is undertaken separately for each of the two components of the wind.

INCOMPRESSIBLE VARIATIONAL ANALYSIS OF WIND FIELD

All approximations applied here, which are not all theoretically fully justified, are sources of mistakes in the initial interpolated wind field. To get a physical constraint on the initially interpolated wind field as described, variational analysis with a strong incompressible constraint is used (Sasaki, 1970; Sherman, 1978; Vrhovec, 1988).

According to Sasaki (1970) and Sherman (1978) the variation formula is written:

$$\iiint_v [\alpha(u - u_i)^2 + \alpha(v - v_i)^2 + \mu(w - w_i)^2 + \lambda \nabla \cdot \vec{u}] \, dxdydz \tag{5}$$

The function is minimized in whole model space V. The velocity components, with indexes i, are initial interpolated values while those without indexes are final, adjusted, values. Parameters α and μ determine the proportion of adjustment of the horizontal and vertical components of velocity, while the Lagrange multiplier λ is a measure for incompressibility request. After some algebra, the functional transforms into a Euler-Lagrange equation that is solved, with appropriate boundary conditions, with finite differences on a fully staggered grid.

POLLUTION TRANSPORT, DISPERSION AND DEPOSITION

The three-dimensional wind field obtained by the procedure described above supplies the meteorological background for a model that simulates the advection, the dispersion and the deposition of an air pollutant.

In the case of headwater areas in NW Slovenia (Upper Sava river valley and Ljubljana basin) the most important air pollutants are SO_2 and sulphates.

To determine the spatial distribution of the pollutants at the ground surface, a three-dimensional time-dependent model was constructed.

$$\frac{\partial C}{\partial t} = - u \frac{\partial C}{\partial x} - v \frac{\partial C}{\partial y} - w \frac{\partial C}{\partial z} - W_d \frac{\partial C}{\partial z} - K \frac{\partial^2 C}{\partial z^2} + S \qquad (6)$$

The basic equation of the model 6 describes three-dimensional air pollutant advection (first three terms on the right), advection with deposition velocity (fourth term), turbulent dispersion (fifth term). Sixth term S describes the sum of all inputs, outputs and transformation of the air pollutant (also deposition at the ground) while the left side of the equation 6 is local time change of the air pollutant concentration. The parameter K represents the turbulent diffusivity and is determined from the wind and temperature field. The parameter W_d is dry deposition velocity. This acts as an additional vertical advection in the atmosphere and enables deposition at the ground.

Some special attention is given to the sixth term S in equation 6. If chemical inactivity of the air pollutant is assumed (for example: sulphates) the only way to remove the pollutant is to deposit it on the ground. It is further assumed that the advected air, at all the boundaries of model space, has constant properties throughout the simulation and that the air above the upper model boundary has the same concentration as that at the upper boundary.

Dry and wet deposition were considered in our simulation. In the case of dry deposition, the pollutant is removed from the lowest layer of the atmosphere by the dry deposition velocity only, in wet case the pollutant is removed from the atmosphere below the mixing height by precipitation scavenging. Both mechanisms of deposition are parametrised in the presented model.

Dry deposition mass flux:

$$\Phi_{mD} = C^1 w_d \qquad (7)$$

and dry deposition sink term at the lowest level (with index 1)

$$S^1 = \frac{\Phi_{mD}}{\Delta z^1} \tag{8}$$

Here Δz^1 denotes the height difference between two lowest levels.

Wet deposition mass flux consists of 'dry' and 'wet' term:

$$\Phi^k_{mW} = C^1 w_d \, \delta^{1,k} + K^k_w \, C^k \, \Delta z^k \tag{9}$$

where Φ^k_{mW} is mass flux from the level with index k, $\delta^{1,k}$ is Kronecker symbol and K^k_w is scavenging efficiency for level with index k. The total deposition at the ground is the sum of all Φ^k_{mW} between the ground and the mixing height.

Wet deposition sink terms are evaluated for all the levels below mixing height.

$$S^k = \frac{\Phi^k_{mW}}{\Delta z^k} \tag{10}$$

The term K^k_w is determined from precipitation rate P:

$$K^k_w = WP/\Delta^k \tag{11}$$

where W is scavenging ratio taken to be 2×10^5, while precipitation rate is assumed to be proportional to vertical velocity of the rising air in saturated atmosphere.

CASE STUDY: UPPER SAVA VALLEY

To test the performance of the coupled models a case study was made. The test area is in part of NW Slovenia on the southern side of the Alps. The Sava headwater area consists of the main Sava valley (with two forks: Sava Dolinka and Sava Bohinjka) and several tributaries (Sora valley, Tržiska Bistrica, Radoyna). The headwater area ranges in altitude from 450 to 2500 metres a.m.s.l. In the model, topography is represented in geographic grid (N-S: 1.87' = 3470 m, W-E: 2.5' = 3200 m). The whole area is 58 × 38 km (18 × 12 grid points in horizontal) and there are 18 levels from the lowest point (ca 450 m a.m.s.l.) to 5000 metres.

The wind data were measured on August 12, 1992, 12.00 UTC at conventional (synop and u.a. stations). The initially-interpolated wind field was adjusted to the constraints and the resulting horizontal wind is shown ˜s Fig. 3 (A) at height 1415 m above mean sea level. The resulting wind is incompressible, but its correspondence to measured values at the ground still strongly depends on the success of the initial interpolation. Though data are measured for horizontal winds, variational adjustment with the incompressibility constraint procedure supplies the vertical component of the wind field (see Fig. 3 (B)).

The initial air pollutant concentration field was assumed to be constant. Above the upper model boundary (and all lateral boundaries), the air has the same concentration

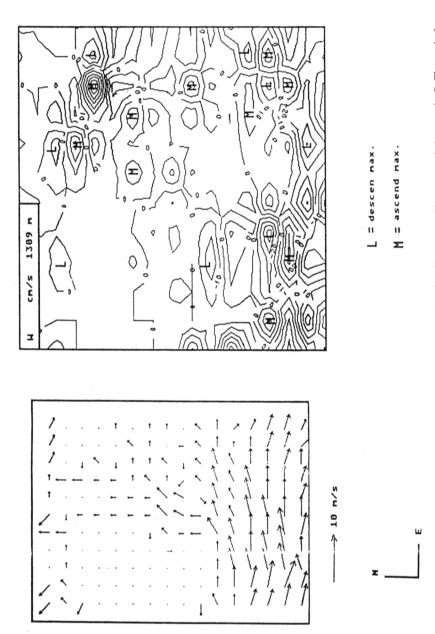

Fig. 3: A. The wind field above the area of analysis at 1415 m above mean sea level after the adjustment. White areas are below terrain. B. The vertical component of the wind field at 1309 m a.m.s.l.

of pollutant. Our first experiment studies dry deposition from homogeneously polluted air with a realistic wind pattern above a real complex topography.

After the integration of the dispersion model for 12 hours, air pollutant concentration had changed and some air pollutant was deposited at the ground. Final air pollutant concentrations at 785 and 1415 m a.m.s.l. are presented of Figs. 4 A and B and the mass of the deposited pollutant is shown on Fig. 5.

The same experiment was repeated with parametrisation for wet deposition as described in equation 9. For this case also the mass deposited at the ground in 12 hours (Fig. 6) is shown. Our second experiment studies dry and wet deposition from homogeneously polluted air on a realistic wind field above real, complex, topography.

The initial concentration was taken to be uniformly 0.1 mg/kg. After 12 hours of integration, significant changes happened. The concentrations in the atmosphere near the ground were reduced due to deposition (in dry case, Fig. 4), while in some regions of the atmosphere (where convergence of vertical velocity is found) the concentrations increased up to three times. The deposited mass of the pollutant follows the pattern of the topography, the vertical velocity distribution and the pattern of the air pollution concentration in the same atmosphere. In the dry deposition case (Fig. 5) the deposition pattern is chaotic and there are very big differences in the deposit over short distances because deposition is mostly controlled by local ascends and descends of the air flow. There are local maxima on all windward sides of the dominant mountain ridges and the deposit is big also on the elevated plateaux. In the wet case, the deposit (Fig. 6) is much more regular as there is predominant forcing of deposition by scavenging with precipitation. When precipitation effects are combined with dry deposition, the pattern shows increase in the deposition along the direction of the wind. The total deposit is, in the wet case, up to 25 times greater than in dry case. Also in wet case, the spatial variability of the deposit is great.

CONCLUSION

- Our hypothesis that in a complex topography with realistic wind field, the distribution of air pollutant deposits is very variable, proved to be correct.
- As wet and dry deposition are both controlled by very irregular vertical velocity distribution (the irregularity comes from the complexity of the topography and from the incompressibility of the flow) the deposit distribution in the area coincides with the main vertical velocity features.
- In the case of wet deposition, the small-scale variability of the deposit is smaller than in dry case. Here spatial variability can be up to 25 times greater in wet case rather than in dry case deposition. The special case, where deposition was calculated using an initial homogenous distribution of the concentration of air pollutant in the atmosphere (like deposition from background pollution) results should be close to reality.

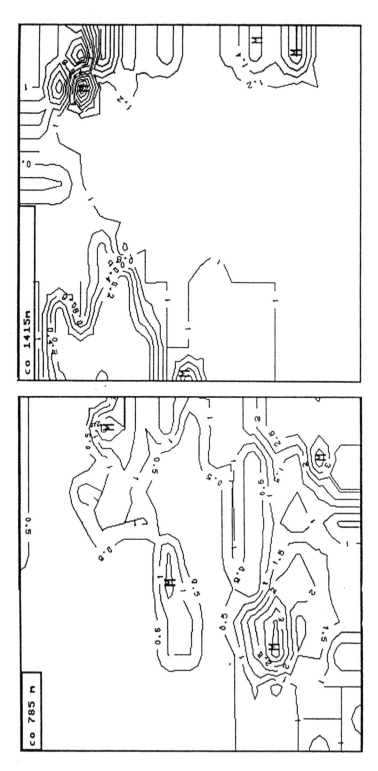

Fig. 4: A. The air pollutant concentration field above the area of analysis. A: at 785 m a.m.s.l.; B. at 1415 m a.m.s.l.

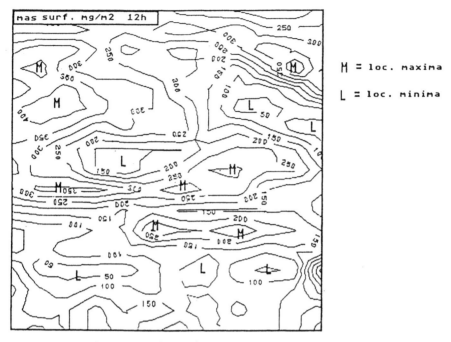

Fig. 5: A. The mass of the deposited air pollutant at the ground on the area of analysis.

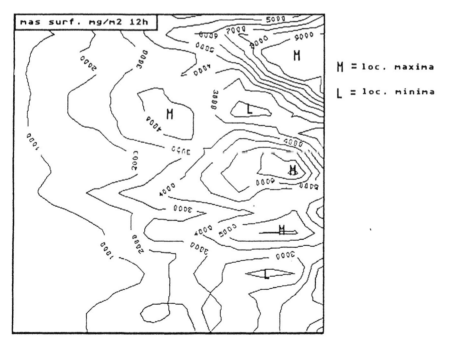

Fig. 6: A. Wet case: The mass of the deposited air pollutant at the ground on the area of analysis.

- If realistic, unhomogenous sources and therefore highly heterogenous concentrations of air pollutants in the atmosphere were assumed, deposition patterns would be even more complex.
- When planning air pollution monitoring networks in a complex topography it is very important (as shown with our model study) where to consider each site for the measuring stations. A very small displacement of the station can produce very big differences in measured values.
- A simulation model of such as that presented, which use topographical features of the area and the general circulation of the atmosphere over the area, can be used to guide quantitative assessments of air pollution deposition distributions.

REFERENCES

Andersson, E., Gustafsson, N., Meuller, L. and Omstedt, G. 1986. Development of meso-scale analysis schemes for nowcasting and very short-range forecasting. *SHMI Promis-rapporter* 1, Norrkoeping, 32 pp.

Anonymous, 1992. *The Environment in Europe: A Global Perspective.* RIVM project 481505, 119 pp.

ApSimon, H.M., Goddard, A.J.H., and Wrigley, J. 1985. Long-range atmospheric dispersion of radioisotopes—I. MESOS model. *Atmos. Environment* 19: 99–111.

ApSimon, H.M., Kruse, M. and Bell, J. 1987. Ammonia emissions and their role in acid deposition. *Atmos. Environment* 21: 1939–1956.

Bosanquet, C.H., Carey, W.F., and Halton, E.M. 1950. *Proc. Inst. Mech. Engrs.* London, 162, 355 pp.

CCU, 1973. *The Long Range Transport of Air Pollutants.* Norw. Ins. for Air Research, Kjeller.

Hrček, D. *et al.*, 1988. *Proučitev mezoklimatskih razmer v občini Velenje.* Občinska raziskovalna skupnost Velenje. No. P -/87 - ORS, Hidrometeorološki zavod Slovenije, Ljubljana, 126 pp.

Klaic, Z. 1990. A Lagrangian one-layer model of long range transport of SO_2. *Atmos. Environment* 24A (7): 1861–1867.

Melgarejo, J.W., and Deardorff, J.W. 1974. Stability functions for the boundary-layer resistance laws based upon observed boundary-layer heights. *J. Atmos. Sci.* 31: 1324–1333.

NILU, 1984. *Summary Report from Chemical Coordinating Centre for Second Phase of EMEP.* NILU-CCC-EMEP 2/84.

OECD. 1974. *Co-operative Technical Programme to Meassure the Long-Range Transport of Air Pollutants.* Nr/ENV/74.7, Paris.

OECD, 1979. *The OECD Programme on Long-Range Transport of Air Pollutants.* Measurements and findings, OECD, Paris.

Petkovšek, Z. 1974. Ocena transkontinentalnega transporta onesnaženja zraka v Slovenijo in iz nje. *Razprave-Papers* XVII, Društvo meteorologov Slovenije, Ljubljana, 11–28B.

Rakovec, J. in Z. Petkovšek, 1975. Približno dolo anje višine nizkih in srednje visokih dimnikov v Sloveniji. *Razprave-Papers* XVIII, Društvo meteorologov Slovenije, Ljubljana 43–62.

Rakovec, J. 1983. The PBL model based on closure hypothesis and with predicted temperature at the ground. *Arch. Met. Geoph. Biocl. Ser.* A, 32: 257–267.

Rakovec, J., Vrhovec, T. and Pristov, N. 1992. Mesoscale variational analysis of wind field in hilly terrain. *CIMA-92, Toulouse, France, Proceedings,* 172–176.

Sasaki, Y. 1958. An objective analysis based on the variational method. *J. Met. Soc. Japan* 36: 77–78.

Sasaki, Y. 1970. Some basic formalisms in numerical variational analysis. *Mon. Weath. Rev.* 98: 875–883.

Sherman, C.A. 1978. A mass consistent model for wind fields over complex terrain. *J. Appl. Met.* 17: 312–319.

Sutton, M.A., Moncrieff, J.B. and Fowler, D. 1992. Deposition of atmosphere ammonia to moorlands. *Environ. Pollut.* 75: 15–24.

Tennekes, H. 1973. Similarity laws and scale relations in planetary boundary layers. *Workshop on Micrometeorology* (D.A. Haugen, ed.), 1777–216 Boston, American Meteorol. Soc., xii + 392 pp.

UKRGAR (United Kingdom Review Group on Acid Rain), 1990. *Acid Deposition in the United Kingdom. 1986–1988.* Warren Spring Laboratory, UK.

Vrhovec, T. 1988. Analysis of three-dimensional mesometeorological wind fields. *CIMA 88, Sestola 18–25 Sept. 1988, Proceedings,* Vol. 2.

Vrhovec, T. 1988a. Mezometeorološka analiza temperaturnih in vetrovnih polj, University Ljubljana, Dept. Physics, M.Sc. Thesis, 84 pp.

Završek, J. 1989. Numerični model nestacionarne difuzije v razgibanem reliefu, University Ljubljana, Dept. Physics, B.Sc. Thesis, 58 pp.

Zilitinkevich, S.S. 1975. Resistance laws and prediction equations for the depths of the planetary boundary layer. *J. Atmos. Sci.* 32: 741–752.

Two Headwater Lakes, a Centre of Research for Long Term Monitoring in Kejimkujik National Park, Nova Scotia, Canada

Joseph J. Kerekes

ABSTRACT

Comparison of two, nearly identical lakes in the remote headwater region of Kejimkujik National Park, Nova Scotia, was undertaken to determine the impacts of the long-range transportation of air pollutants. The two lakes, located in a pristine environment, represent the most sensitive receptor systems in North America. Since the later 1970s, they have been the focus for a multi-agency Canadian Government study. This paper reports a base-line ecological characterization of the lakes.

Keywords: Headwater lakes, Kejimkujik National Park, air pollutants, Canada.

INTRODUCTION

In the past decades, there has been a growing scientific, popular and political concern over environmental damages caused by various human activities. In the late 1970s the Government of Canada initiated the Kejimkujik Calibrated Study Program in response to the need for scientific knowledge of the potential impacts of the long-range transportation of air pollutants (LRTAP) on terrestrial and freshwater ecosystems in Canada. This is a multiagency study of numerous federal government agencies, as well as contracted scientists from universities and the private sector. The program has been inter-disciplinary, with a wide array of studies among the biological, chemical and atmospheric sciences (Kerekes, 1989a, b; Kerekes and Freedman, 1989).

The study area and watersheds have particular advantages for such a research program. These include: i) their location in a pristine and undisturbed environment, remote from local anthropogenic emissions of air pollutants, and presumably exposed only to natural deposition and LRTAP; ii) their protection within a national park; iii) their weakly-buffered oligotrophic watersheds, and the small bicarbonate buffering capacity of their lakewater (and hence sensitivity to acidification); iv) the two lake basins represent the most sensitive receptor systems in North America (calcium concentration approx. 20 µeq/l); v) of the two headwater lakes selected, one has clear water while the other is coloured (influenced by organic acids). The two lakes have nearly identical water chemistry except for organic acidity. The pH of the coloured lake is about one unit lower, pH 4.4 vs. 5.3 in the clear lake. Thus are ideal to evaluate the relative importance of natural (organic) vs. anthropogenic acidity (sulfates, nitrates) on sensitive surface waters; vi) the watersheds receive maritime dominated precipitation high in Na and Cl. Snow and rain alternate in the winter, in contrast to continental locations in Canada; vii) the availability of baseline data on meteorology, hydrology, water chemistry, aquatic biology, and terrestrial biophysical characteristics; and viii) the study region receives acidic deposition, and there is evidence that some surface waters have acidified (Shaw, 1979; Watt et al., 1979, 1983; Kerekes, 1980; Thompson et al., 1980; Elder and Martin, 1989) and that Atlantic Salmon (*Salmo salar*) have disappeared from some acidic rivers (Farmer et al., 1980; Watt et al., 1983).

The integrated research program began in 1978 with an intensive baseline characterization phase between 1980 and 1982. Some studies have continued to the present to provide a longer-term data base, most notably monitoring of the chemistry and hydrology of precipitation and surface water, and various components of the aquatic biota such as invertebrates, fish and piscivorous birds. The purpose of this report is to summarize and synthesize the chemical and biological information of two oligotrophic, acidic headwater lakes which formed the cornerstone of the Kejimkujik Study.

STUDY AREA

The two headwater lakes lie within Kejimkujik National Park, in southwestern Nova Scotia, Canada. The park is 94% forested with about 75% of the forest comprised mixed angiosperm conifer stands, 20% coniferous, and 4% angiosperm (Gimbarzevsky, 1975).

Kejimkujik National Park has at least 46 lakes and ponds, the largest of which is Kejimkujik Lake, 25 km^2 in area (Kerekes and Schwinghamer, 1973). All of these water bodies are oligotrophic, and many of them are tea-coloured because of a large concentration of dissolved organic substances that have leached from boggy, organic substrate that has developed on poorly-drained and undrained sites within their watershed (Kerekes, 1975a).

Study Watersheds

The location and morphometrics of the two headwater lakes, Beaverskin and Pebbleloggitch are summarized in Table 1. The lakes are about 60 km inland from the Atlantic Ocean and 50 km from the Bay of Fundy. The ratio of drainage basin/lake surface area is 2.3 in Beaverskin Lake, and 4.7 in Pebbleloggitch Lake. The lakes are oligotrophic, Pebbleloggitch Lake has brown-coloured water because of a large concentration of soluble organic substances, primarily fluvic acids (Gjessing, 1976),

Table 1: The location and selected morphometrics of the two study lakes. Modified from Kerekes *et al.* (1982)

Variable	Beaverskin	Pebbleloggitch
Location	44° 18′N	44° 18′N
	65° 20′W	65° 21′W
Elevation (m)	120	120
Lake area (km²)	0.443	0.338
Drainage basin +lake (km²)	1.0	1.6
Maximum depth (m)	6.8	2.5
Mean depth (m)	2.8	1.6
Water retention time (yr)		
June 1978 to May 1980	1.36	0.38

leached from the organic substrate of heath and *Sphagnum*-dominated wetlands, which are prominent in their watersheds.

The lakes do not have a permanent surface inflow. Most of their hydrologic input from the terrestrial watershed takes place by subsurface, near-shore seepage of groundwater that flows over the bedrock (Jacques *et al.*, 1982, 1983).

The bottom substrate within 10–20 m. of the shore of the lakes in dominated by a coarse and rocky till, and erratic boulders are frequent. Interstices have a fine matrix that is variously comprised of sand, silt, clay and organic material. Beyond this near-shore zone, the sediment is uniformly fine-grained, with mixtures or organic, silt, and clay minerals, and occasional projections of large erratics (Kerekes and Freedman, 1989).

Terrestrial Watersheds

Bedrock of the watershed of Beaverskin Lake is all slate, while the watershed of Pebbleoggitch Lake contains granite and slate (Gimbatrzevsky, 1975; EESL, 1976).

Soil of the terrestrial watershed of Beaverskin and Pebbleloggitch lakes is largely mapped in the Gibraltar Series. However, about 1/3 of the Pebbleloggitch watershed has a peaty organic substrate developed under a heath-*Sphagnum* plant community on undrained and poorly drained sites (Gimbatrzevsky, 1975; Percy and Kerekes, 1981). Seismic and other surveys by Jacques *et al.* (1982) indicated that the depth of till over bedrock was 0–7 m in the Beaverskin watershed. Gibraltar soil is notably deficient in calcium frequently containing < 1000 kg/ha of total calcium within the tree rooting depth (Freedman and Morash, 1985). The paucity of calcium in the terrestrial watershed is important with respect to susceptibility to acidification since the availability of calcium carbonate is an important edaphic/bedrock factor that influences the buffering and acid-neutralizing capacity of soil and water in the circumneutral pH range (Kramer, 1978; Henriksen, 1980).

WATER QUALITY

The ionic composition of water in the study lakes is summarized in Table 2 and Figure 1. The lakes are dominated by sodium and chloride, which reflects a marine influence. Some obvious differences in other chemical constituents can be attributed to differences in the drainage basins of the lakes. Of particular importance is a paucity of dissolved organic substances in Beaverskin Lake, as indicated by a relatively low water colour (5 Hazen units) and a small concentration of organic anions. In contrast Pebbleloggitch Lake received drainage from peaty wetlands, and has a relatively intense colour (118 Hazen units) and a large concentration of organic anions. The water of the lakes is dilute (the sum of inorganic constituents is about 11 mg/L). The concentration of dissolved organic carbon is sufficiently large in the coloured lake (11.9 mg/L) it exceeds the sum of inorganic constituents. Kerekes *et al.* (1986) suggested that although organic acidity was important in organic water in the study area, sulfate also contributes to acidity, particularly during the time of high water discharge in the spring and autumn.

Table 2: Average chemical composition of two lakes based on monthly mean values in Kejimkujik National Park, Nova Scotia for the period of June 1982 to May 1983 (modified from Kerekes *et al.*, 1986)

	Beaverskin	Pebbleloggitch
	(μeq/l)	
Ca	18.7	19.3
Mg	29.1	30.9
Na	114.5	123.1
K	6.7	7.0
H	5.3	46.2
Fe	3.0	9.3
Al	5.6	26.0
NH_4	2.8	5.0
$NO_2 + NO_2 +$ (as NO_3)	0.2	0.2
SO_4	43.4	45.2
Cl	116.1	112.1
Organic anion	26.7	99.2
Gran alkalinity	2.0	−25.5
Sum of constituents	374.1	523.5
pH	5.3	4.3
Color (Hazen u.)	5.1	118.0
Dissolved organic carbon (mg/l)	3.3	16.3

The concentration of sodium and chloride is relatively high in both Beaverskin and Pebbleloggitch (112–123 μeq/L). The watersheds of Beaverskin and Pebbleloggitch lakes (both 19 μeq Ca/L) lie on igneous bedrock with shallow noncalcareous till and the calcium concentration in these lakes is amongst the smallest in the world (Armstrong and Schindler, 1971; Kerekes *et al.*, 1982). Magnesium concentration

Beaverskin Lake

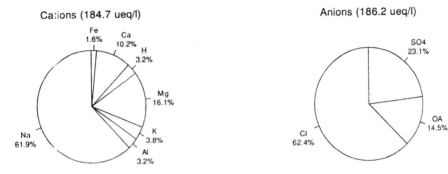

Pebbleloggitch Lake

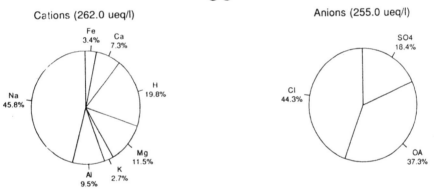

Fig. 1: Average chemical and ionic composition of Beaverskin and Pebbleloggitch lakes.

(30–39 µeq/L) is also relatively small. The paucity of calcium and magnesium indicates that these surface waters are sensitive to acidification.

Total aluminum and iron are present in a relatively large concentration in the coloured lake. The total aluminum concentration (26 µeq/L) exceeds that of calcium in Pebbleloggitch. Nitrate and ammonium concentration is small in the two lakes, particularly during the growing season when they average < 1 µeq/L. Because inorganic nitrogen is present in a relatively large concentration in precipitation (Kerekes and Freedman, 1989), the small concentration in surface water suggests that it is removed from water by contact with vegetation and soil of the terrestrial watershed (Kerekes et al., 1982). Therefore, under existing atmospheric loading rates, nitrate may not contribute directly to the acidification of surface water in the study area; rather, it acts as a fertilizer to the terrestrial vegetation.

The average annual pH of 5.3 in the clearwater Beaverskin Lake indicates that the relatively modest atmospheric loading of non-marine sulfate of about 14 kg/ha yr

(Beattie, 1993) is sufficient to maintain moderate acidity in a weakly-buffered lake. The effect of organic acidity is best seen in Pebbleloggitch Lake (pH 4.3). Beaverskin and Pebbleloggitch have a nearly identical calcium concentration, but the concentration of organic anions is considerably large in the coloured Pebbleloggitch Lake. As a consequence, the average H+ concentration (46 μeq/L. or 20% of cations) is substantially larger in Pebbleloggitch than in Beaverskin (5.3 μeq/L. or 3% of cations).

The differences in acidity between the lakes is reflected in their Gran alkalinity. Beaverskin has near-zero alkalinity throughout the year (annual average = 2.0 μeq/L), while Pebbleloggitch has negative alkalinity (annual average = −26 μeq/L).

Phosphorus, Chlorophyll, and Trophic State

The annual mean concentration of chlorophyll-a and total phosphorus was higher in Pebbleloggitch Lake (Table 3). The data for annual mean chlorophyll-a and total phosphorus for Beaverskin plot close to the oligotrophic end of the log-log relationship of these variables for a large sample of temperate lakes (Fig. 2), indicating that this lake has a typical trophic response to phosphorus. The coloured lake falls substantially beneath the line, which is a typical response for a coloured lake in which only a small proportion of the total dissolved phosphorus is biologically available (Kerekes, 1983b).

Table 3: Miscellaneous chemical constituents in water of the study lakes. Arithmetic mean values of eight depth-integrated samplings between June 9 and Sept. 23, 1981 (modified from Blouin *et al.*, 1984)

Variable	Beaverskin	Pebbleloggitch
Chlorophyll-a (mg/m³)	0.67	1.60
Phaeophytin ((mg/m³)	0.90	2.70
Dissolved inorganic carbon (mg/L)	0.34	0.46
Total organic carbon (mg/L)	4.20	16.5
Dissolved organic carbon (mg/L)	3.60	11.9
Total P (mg/m³)	5.40	13.6
Total dissolved P (mg/m³)	3.30	10.5
Soluble reactive (mg/m³)	1.10	2.20
Total N (mg/L)	0.20	0.32
Ammonium (mg/L)	0.05	0.09
Nitrate + nitrite (mg/L)	0.01	0.01
Dissolved O_2	7.80	8.10
% oxygen saturation	92.00	93.00
Zooplankton biomass, 1981 (mg/m³)	151.00	243.00

Although the coloured lakes are relatively acidic, hydrogen ion should not in itself be considered to be a factor that depresses the trophic response. Kerekes *et al.* (1984) showed that a very acidic lake (pH 3.6) in Nova Scotia had a typical phosphorus to chlorophyll response, while a nearby acidic (pH 4.0) eutrophic lake in which a large

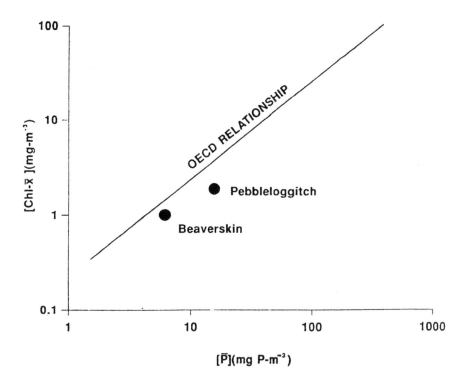

Fig. 2: Annual mean chlorophyll and total phosphorus relationships between Beaverskin and Pebbleloggitch lakes.

proportion of the total phosphorus was biologically available, plotted well above the chlorophyll vs. total phosphorus line.

The lakes are oligotrophic with respect to their chlorophyll concentration (Table 3), according to a proposed boundary for this trophic condition (i.e. annual mean chlorophyll ≤ 2.5 mg/m^3. Vollenweider and Kerekes, 1981a, 1981b). With respect to total-phosphorus, Beaverskin is oligotrophic (i.e. < 10 mg/m^3), while Pebbleloggitch is borderline mesotrophic (i.e. 10–35 mg/m^3, Vollenweider and Kerekes, 1981a, 1981b). However, because only a small proportion of the total-P is biologically available in the coloured lakes, they and all other lakes in Kejimkujik National Park should be considered as oligotrophic (Kerekes, 1975a).

Phytoplankton and Planktonic Primary Production

Annual planktonic primary production was higher in the clear water Beaverskin Lake, as was production averaged over the ice-free season (Table 4). The difference in unit-area primary production between lakes is partly related to a difference in transparency (extinction coefficient in Beaverskin = 0.44/m and Pebbleloggitch = 2.2/m: see also Figure 3). Pebbleloggitch Lake had the smaller unit-volume productivity, but it had the larger rate of volumetric productivity at light optimum between lakes. The latter effect may be related to the relatively large concentration of total phosphorus, total dissolved phosphorus, and soluble reactive phosphorus during the growing season in Pebbleloggitch Lake (Table 3). The relatively large concentration of phosphorus could have been due to several factors: i) the mixing layer is much deeper than the euphotic zone in Pebbleloggitch Lake, so there is an overturn of relatively P-rich water during windy conditions; and ii) the decomposition of dissolved and suspended allochthonous organic matter in the water column may release nutrients such as phosphorus, and may sustain heterotrophic production by planktonic algae (Ilmavirta, 1980, 1982). The relatively small primary production under field conditions in Pebbleloggitch is caused by a restricted transparency of the water column and a consequently shallow euphotic zone rather than by its acidity.

Table 4: Annual planktonic primary production, mean ice-free season areal production and productivity in two acid stressed lakes (modified from Beauchamp, 1983). Units of primary production after Kerekes (1975)

1980–81	Beaverskin	Pebbleloggitch
Annual production		
P-area (gC m^{-2} yr^{-1})	82.8	34.4
P-area $\bar{x}$ (gC m^{-2} yr^{-1})	52.8	29.1
Ice-free Season Mean		
P-area (mgC m^{-2} d^{-1})	156.6	62.5
P-area $\bar{x}$ (mgC m^{-2} d^{-1})	97.6	59.1
P-max (mgC m^{-3} d^{-1})	46.9	72.8
P-vol $\bar{x}$ (mgC m^{-3} d^{-1})	34.8	36.9

Howell and Kerekes (1987) showed that in lakes where the ratio of z:z_{eu} (mean depth/euphotic zone depth) is much less than unity, P-area production estimates from the deep station will generally overestimate aerial production, as the euphotic zone is restricted due to the shallowness of the lake. To overcome this, Kerekes (1975c) proposed the use of a lake basin volume corrected P-area value which he termed P-area $\bar{X}$. Fee (1979) observed that use of P-area overestimation actual lake production by approximately 20%. However, as this overestimation of production is a function of both basin morphometry and transparency, clearwater lakes which have a large littoral zone and a small area of deep water (i.e., low z:z_{cu}), may have errors much greater than 20%. In the clear water Beaverskin Lake, this error of overestimation is about 60%, while in the highly coloured, very shallow Pebbleloggitch Lake P-area $\bar{X}$ production overestimates P-area production by 25% (Fig. 4).

Light extinction curves July 8, 1979

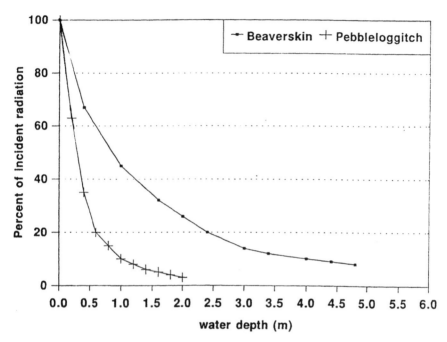

Fig. 3: Light extinction curves for Beaverskin and Pebbleloggitch lakes.

Phytoplankton cell density in Beaverskin Lake was dominated by Blue-green algae (94% of total cell density), followed by Greens (2.9%). Chrysophyceans (1.4%), and Euglenoids (1.3%) (Blouin *et al.*, 1984). Kwiatkowski and Roff (1976) found that blue-green algae dominated the phytoplankton of clearwater acidic lakes near Sudbury, Ontario, as we observed in clearwater Beaverskin Lake. The much smaller cell density in Pebblologgitch lake was dominated by Chrysophyceans (32% of total cell density), Greens (31%), Diatoms (17%), and Zanthophyceans (17%). The dominant phytoplankton groups in Pebbleloggitch were Diatoms contributing an average 77% of the total cell density, Greens 9%, and Chrysophyceans 6% (Blouin *et al.*, 1984). These families have also been prominent in studies of the algal flora of coloured water bodies elsewhere (Ostrovsky and Duthie, 1975; Ilmavirta, 1980; Nakatsu, 1983).

In Beaverskin Lake, Green algae were relatively diverse (average of 15 taxa), and the most abundant species was *Sphaerocystis schroeteri*, Blue-green algae were dominated by *Agmenellum thermale* (Blouin *et al,*, 1984). The smaller importance of Blue-green algae in the relatively coloured and acidic lakes, may be due to the general intolerance of these algae to acidity (Block, 1973). In Pebbleloggitch Lake, there was a relatively large species richness of Green algae and Diatoms. The most abundant

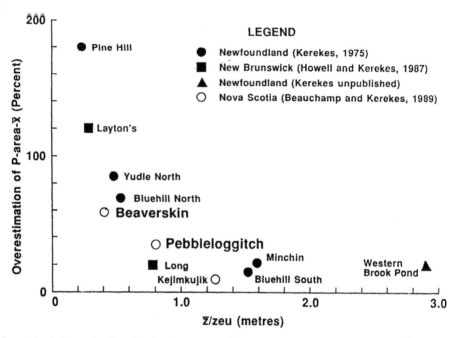

Fig. 4: Overestimation of basin volume P-area-$\overline{X}$ and mean depth/euphotic zone depth ($\overline{Z}/Z_{eu}$) relationships for the lakes of study.

phytoplankton species were the Chrysophycean *Mallomonas caudata* in 1980, and the Green alga *Sphaerocystis schroeteri* in 1981. Detailed lists of phytoplankton taxa are in Beauchamp (1983) and Blouin (1985).

Macrophytes

Virtually all of the benthic substrate of Beaverskin Lake was vegetated with macrophytes (after Stewart and Freedman, 1982, unpublished). The most widespread macrophyte community was dominated by *Sphagnum macrophyllum* and *Utricularia vulgaris,* with *U. purpurea* occasionally prominent. This community was present beneath 71% of the lake surface, in a water depth of 2.5–6.5 m. The average cover of *Sphagnum* was 55%.

Most of the benthic substrate of brownwater Pebbleloggitch Lake was an unvegetated, silty bottom with frequent erratic boulders. The most extensive macrophyte community was dominated by *Nuphar variegatum,* and it covered 9% of the bottom, at a depth of <1.5 m. Detailed lists of macrophyte species is given by Stewart and Freedman (1982), Catling *et al.* (1986) and summarized by Kerekes and Freedman (1989).

The difference in macrophyte distribution and standing stock between Beaverskin Lake (average of 610 kg d.w./ha) and Pebbleloggitch Lake (25.5 kg/ha) is due to the difference in their water colour. Light penetrates to the bottom of the clearwater

Beaverskin Lake (Fig. 1), and as a result the benthic substrate of this lake is 96% vegetated and macrophytes are present as deep as 6.5 m. The brownwater Pebbleloggitch Lake is only 15% vegetated, and macrophytes are limited to a depth < 1.5 m.

The invasion of acidified lakes by bryophytes, particularly *Sphagnum* species, may contribute to acidification. Acidophilous *Sphagnum* species have an efficient cation exchange ability at their surface which results in the removal of basis cations such as calcium, magnesium, and potassium from solution, and the release of hydrogen ion in exchange (Skene, 1915; Clymo, 1963, 1964, 1984; Grahn, 1976). The relatively high pH Beaverskin Lake is dominated by a benthic *Sphagnum* community, but this moss was not prominent in the brownwater Pebbleloggitch Lake. It appears that the acid-generating potential of *Sphagnum* has not caused a severe acidification of Beaverskin Lake.

Invertebrates

Zooplankton biomass and density averaged largest in Pebbleloggitch Lake (Table 3). The relatively large biomass in this lake could be related to its large concentration of allochthonous suspended organic matter, a factor which has been hypothesized to sustain a relatively large productivity of zooplankton in other brownwater lakes (Nauwerck, 1963; Schindler and Noven, 1971; Ostrovsky and Duthie, 1975). Zooplankton species and zooplankton density is given by Blouin *et al.* (1984) and summarized by Kerekes and Freedman (1989).

Schell and Kerekes (1989) noted the leathery thinness of the shells of the sheriid clam, *Pisidium casertauum* in Beaverskin Lake. They suggested that pH limits the distribution of this Pelecypod in extremely low Ca lakes of Southwestern Nova Scotia. Gastropods were absent from the acidic study lakes.

Of the 12 fish species of Kejimkujik National Park (Kerekes, 1975, 1982; Kerekes and Freedman, 1989) seven species were found in Beaverskin and only one species, the Yellow Perch (*Perca flavescens* Mitchill) is breeding in Pebbleloggitch. Beamish (1976) reported a reproductive failure of yellow perch at pH 4.5–4.7, while they were present in Pebbleloggitch with seasonal pH values as low as 4.1 (Kerekes, 1975). With respect to amphibians Dale (1984) reports that Beaverskin and Pebbleloggitch Lakes both had adult spring peepers (*Hyla crucifer*), green frog (*Rana clamitans*), pickerel frog (*R. palustris*), bullfrog (*R. catesbeiana*), and American toad (*Bufo americanus*), while Beaverskin also has spotted newt (*Notophthalmus viridescens*), and Pebbleloggitch had leopard frog (*R. pipiens*). No information is available on the breeding success of amphibians in these lakes, but their water chemistry falls well within the range of variation of successful habitat in Nova Scotia (Dale *et al.*, 1985a). In addition, laboratory bioassays indicated that the water of Beaverskin and Pebbleloggitch Lakes could sustain a successful hatching and early embryonic development of several species of amphibian (Dale *et al.*, 1985b). Both common loons (*Gavia immer*) and common mergansers (*Mergus merganser*) are known to breed on Beaverskin Lake (Kerekes *et al.*, 1983). Both of those species are occasional on Pebbleloggitch, but they are not known to breed there. However, a wider survey within the national park has shown that both of these species will breed on darkwater lakes.

FUTURE DIRECTIONS

The value of the baseline characterization (physical, chemical and biological) and the monitoring of these in the two headwater lakes and adjacent drainage basins has been recognized. A high degree of cooperation among many agencies developed an infrastructure necessary for long-term monitoring observed (e.g. Kerekes *et al.*, 1986; Allen *et al.*, 1993). Based on this recognition, the Kejimkujik area was selected as the first site of the national, long-term environmental monitoring and assessment of ecosystem components network in Canada, representative of the Atlantic Maritime Ecozone (Kerekes, 1993; St. Pierre, 1993).

REFERENCES

Allen, Y., Clair, T.A., Drysdale, C., Hirvonen, H. and Kerekes, J. 1993. Ecological monitoring and research at Kejimkujik National Park, 1978–1992. *Reg. Env. Monitoring and Res. Coord. Com. Oc. Rep.* No. 2. Env. Canada, 57 pp.

Amstrong, F.A. and Schindler, D.W. 1971. Preliminary chemical characteristics of waters in the Experimental Lakes Area, northwestern Ontario *J. Fish. Res. Board Can.* 28: 171–187.

Beamish, R.J. 1976. Acidification of lakes in Canada by acid precipitation and the resulting effects on fishes. *Water, Air and Soil Pollut.* 6: 501–514.

Beattie, B. 1993. *Report of the Atlantic Region LRTAP Monitoring and Effects Working Group for 1993.* Env. Canada AES Bedford, N.S., 184 pp.

Beauchamp, S. 1983. Planktonic primary production in three acid stressed lakes in Nova Scotia. MSc Thesis. Department of Biology, Dalhousie University, Halifax, N.S., Canada, 194 pp.

Blouim, A.C., Land, P.A., Collins, T.M. and Kerekes, J.J. 1984. Comparison of plankton-water chemistry relationships in three acid stressed lakes in Nova Scotia. *Canada. Int. Revue ges. Hydrobiol.* 69: 819–841.

Blouin, A.C. 1985. Comparative patterns of plankton communities under different regimes of pH in Nova Scotia, Canada. PhD Thesis. Department of Biology. Dalhousie University, Halifax, N.S., Canada, 276 pp.

Brock, T. 1973. Lower pH limit for the existence of blue-green algae: Evolutionary and ecological implications. *Science* 179: 480–482.

Catling, P.M., Freedman, B., Steward, C., Kerekes, J.J. and Lefkovith, L.P. 1986. Aquatic plants of Kejimkujik National Park. Nova Scotia: Floristic composition and relation to water chemistry. *Can. J. Bot.* 64: 724–729.

Clymo, R.S. 1963. Ion exchange in *Sphagnum* and its relation to bog ecology. *Ann. Bot.* 27: 309–324.

Clymo, R.S. 1964. The origin of acidity in *Sphagnum* bogs. *Bryologist* 67: 427–431.

Clymo, R.S. 1984. *Sphagnum*-dominated peat bogs: A naturally acidic ecosystem. *Phil. Trans. T. Soc. Lond.* 305: 487–499.

Dale, J.M. 1984. Effects of acidity and associated water chemistry variables on the occurrence and reproduction of Nova Scotian amphibians. MSc Thesis. Department of Biology. Dalhousie University, Halifax, N.S., Canada.

Dale, J.M., Freedman, B. and Kerekes, J. 1985a. Acidity and associated water chemistry of amphibian habitats in Nova Scotia. *Can. J. Zool.* 63: 97–105.

Dale, J.M., Freedman, B. and Kerekes, J. 1985b. Experimental studies of the effects of acidity and associated water chemistry on amphibians. *Proc. N.S. Inst. Sci.* 35: 35–54.

EESL. 1976. *Kejimkujik National Park.* Resource atlas and base description. Eastern Ecological Services Ltd. Roport on File with Parks Canada. Halifax, N.S., Canada.

Elder, F.C. and Martin, H.C. 1989. Kejimkujik Park-One in a family of integrated watershed studies. *Water, Air and Soil Pollut.* 46: 1–12.

Farmer, G.T., Goff, T.R., Ashfield, D. and Samang, H.S. 1980. Some effects of the acidification of Atlantic salmon rivers in Nova Scotia. *Tech. Rep. Fish Aquatic. Sci.* No. 972. Department of Fisheries. St. Andrews, N.B., Canada.

Fee, E.J. 1979. A relation between lake morphology and primary productivity and its use in interpreting whole-lake eutrophication experiments. *Limnol. Oceanogr.* 24: 401–416.

Freedman, B. and Morash, R. 1985. Forest biomass and nutrient studies in central Nova Scotia. Part 4. Biomass and nutrients in ground vegetation, forest floor, soil and litterfall in a variety of forest stands. *Contractors Report to Canadian Forestry Service.* Fredericton, N.B., Canada.

Gimbarzevsky, P. 1975. *Biophysical survey of Kejimkujik National Park, Information Report* FMR-X-81. Forest Management Institute. Petawawa, Ont., Canada.

Gjessing, E.T. 1976. Physical and chemical characteristics of aquatic humus. *Ann. Arbor. Science Pub. Inc.*, Ann Arbor, MI.

Grahn, O. 1976. Macrophyte succession in Swedish lakes caused by deposition of airborne acid substances. In: *Proc. First Int. Symposium Acid Precipitation and the Forest Ecosystem.* USDA Forest Service. General Technical Report NE-23. Northeastern Forest Experiment Station. Broomall. PA., pp. 519–530.

Henriksen, A. 1980 Acidification of freshwater.—A large-scale titration. In: Drablos, D. and Tollan, A. (eds). *Ecological Impact of Acid Precipitation.* SNSF project. Oslo, Norway, pp. 68–74.

Howell, G. and Kerekes, J. 1987. Primary Production of Two Small Lakes in Atlantic Canada. *Proc. N.S. Inst. Sci.* 37: 71–88.

Ilmavirta, V. 1980. Phytoplankton in 35 Finnish brown-water lakes of different trophic status. pp. 121–130. In: Dukulil, M., Metz, H. and Jensen, D. (eds.). *Developments in Hydrobiology*, Vol. 3. Dr. W. Junk Pub. The Hague, The Netherlands.

Ilmavirta, V. 1982. Dynamics of phytoplankton in Finnish lakes. *Hydrobiologia* 86: 11–20.

Jacques, Whitford, and Associates. Ltd. 1982. Kejimkujik groundwater characterization study. Report to Environment Canada, Project 2281. On file with the Canadian Wildlife Service, c/o Dept. of Biology, Dalhousie University, Halifax, N.S., Canada.

Jacques, Whitford, and Associates. Ltd. 1983. Groundwater characterization. Beaverskin Lake. Kejimkujik National Park. Report to Inland Waters Directorate. Halifax, N.S., Canada.

Kerekes, J. 1975a. Phosphorus supply in undisturbed lakes in Kejimkujik National Park. Nova Scotia, Canada. *Verh. Internat. Verein Limnol.* 19: 349–357.

Kerekes, J. 1975b. *Aquatic Resources Inventory. Part. 7. Distribution of Fishes.* Prepared for Parks Canada by Canadian Wildlife Service, c/o Department of Biology, Dalhousie University, Halifax, N.S., Canada.

Kerekes, J. 1975c. The relationship between primary production and basin morphometry in five small oligotrophic lakes in Terra Nova National Park in Newfoundland. *Symp. Biol. Hungary* 15: 35–48.

Kerekes, J. 1980. Preliminary characterization of three lake basins sensitive to acid precipitation in Nova Scotia, Canada. In: Drablos D. and Tollan, A. (eds.). *Impact of Acid Precipitation.* SNSF Project, Oslo, Norway, pp. 232–234.

Kerekes, J. 1982a. Fish distribution in Kejimkujik National Park. In: J. Kerekes (ed). *Kejimkujik Calibrated Catchments Program.* LRTAP Liaison Office. Atmospheric Environment Service. Downsview, Ont., Canada, pp. 57–61.

Kerekes, J. 1989a. Impact of Acid Rain on Aquatic Systems, The Kejimkujik National Park Study. pp. 141–148. In: Harvey *et al.* (eds), *Use and Management of Aquatic Resources in Canada's National Parks.* Oc. Pap. 11. Heritage Res. Center. U. Waterloo, 262 pp.

Kerekes, J. 1989b. Preface. Acidification of organic waters in Kejimkujik National Park, Nova Scotia. Proceedings of a symposium on the acidification of organic waters in Kejimkujik National Park, Nova Scotia, Canada, held in Wolfville, Nova Scotia, October 25–27, 1988. *Water, Air and Soil Pollut.* 46: vii–viii.

Kerekes, J. 1983b. Predicting trophic response to phosphorus addition in a Cape Breton Island Lake. *Proc. NS. Inst. Sci.* 33: 7–18.

Kerekes, J. and Schwinghamer, P. 1973. Aquatic resources inventory. Kejimkujik National Park, N.S. Part 2. Lake drainage and morphometry. Manuscript Report. Canadian Wildlife service. c/o Department of Biology, Dalhousie University. Halifax, N.S., Canada.

Kerekes, J., Bates, D., Duggan, M. and Tordon, R. 1993. Abundance and distribution of fish-eating birds in Kejimkujik National Park. (1988–1993). *Workshop on the Kejimkujik Watershed Studies: Monitoring and Research, Five Years After "Kejimkujik '88".* Kejimkujik National Park. Oct. 20–21, 1993.

Kerekes, J., Howell, G., Beauchamp, S., and Pollock, T. 1982. Characterization of three lake basins sensitive to acid precipitation in central Nova Scotia (June, 1979 to May 1980). *Int. Revue Ges. Hydrobiol.* 67: 679–694.

Kerekes, J. and Freedman, B. 1989. Characteristics of Three Acidic Lakes in Kejimkujik National Park, N.S., Canada. *Arch. Environ. Contam. Toxicol.* 18: 183–300.

Kerekes, J., Freedman, B., Howell, G., and Clifford, P. 1984. Comparison of the characteristics of an acidic eutrophic and an acidic oligotrophic lake near Halifax. Nova Scotia. *Water Pollut. Res. J. Canada* 19: 1–10.

Kerekes, J., Beauchamp, S., Tordon, R., and Tremblay, C. 1986. Organic versus anthropogenic acidity in tributaries of the Kejimkujik watersheds in western Nova Scotia. *Water, Air and Soil Pollut.* 31: 165–173.

Kramer, J.R. 1978. Acid precipitation. In Nriagu, J.O. (ed.) *Sulfur in the Environment.* John Wiley and Sons. New York, N.Y., pp. 325–370.

Kwiatowski, R.E. and Roff, J.C. 1976. Effects of acidity on the phytoplankton and primary productivity of selected northern Ontario lakes. *Can. J. Bot.* 54: 2546–2561.

Nakatsu, C.H. 1983. The algal flora and ecology of three-brown water systems. MSc Thesis. Department of Botany. University of Toronto, Toronto, Ont., Canada.

Nauwerck, A. 1963. Die beziehungen zqischen zooplankton und phytoplankton. In: *See Erken. Symp. Biol. Uppsala.* 17: 1–163 cited in Ostrovsky and Duthie, 1975.

Ostrovsky, M.L. and Duthie, H.C. 1975. Primary productivity and phytoplankton of lakes on the eastern Canadian shield. *Verh. Internat. Verein Limnol.* 19: 732–738.

Percy, K.E. and Kerekes, J. 1981. Kejimkujik calibrated catchments program on the acquatic and terrestrial effects of the long-range transport of air pollutants. Canadian Wildlife Service, c/o Department of Biology, Dalhousie University. Halifax, N.S., Canada.

Schell, V.A. and Kerekes, J. 1989. Distribution, Abundance and Biomass of Benthic Macroinvertebrates, Relative to pH and Nutrients in Eight lakes of Nova Scotia. *Water, Air and Soil Poll.* 46: 359–374.

Schindler, D.W. and Noven, B. 1971. Vertical distribution and seasonal abundance of zooplankton in two shallow lakes of the Experimental Lakes Area, northwestern Ontario. *J. Fish. Res. Board Can.* 28: 245–256.

Shaw, R.W. 1979. Acid precipitation in Atlantic Canada. *Environ. Sci. Technol.* 13: 366–411.

Skene, M. 1915. The acidity of *Sphagnum* and its relation to chalk and mineral salts. *Annals. Bot.* 29: 65–90.

St. Pierre, R. 1993. *Proc. Atlantic Maritime Ecozone Long-Term Monitoring and Organizational Workshop.* March 30–31, 1993. Halifax, N.S. Reg. Environ. Monitoring Res. Coord. Com. Oc. Rep. No. 1, 43. pp.

Stewart, C. and Freedman, B. 1982. A survey of aquatic macrophytes in three lakes in Kejimkujik National Park. Nova Scotia. Report to Canadian Wildlife Service, c/o Department of Biology, Dalhousie University, Halifax, N.S., Canada.

Thompson, M., Elder, F.C., Davis, A.R., and Whitlow, S. 1980. Evidence of acidification in rivers of eastern Canada. In: Drablos D. and Tollar, A. (eds.) *Ecological Impact of Acid Precipitation*. SNSF Project. Oslo. Norway, pp. 224–245.

Vollenweider, R.A. and Kerekes, J. 1981a. OECD cooperative program on monitoring of inland waters (eutrophication control). Synthesis Report. OECD Secretariat Environmental Directorate. Paris, France.

Vollenweider, R.A. and Kerekes, J. 1981b. Background and summary results of the OECD cooperative program on eutrophication. In: *Restoration of Lakes and Inland Waters*. Pub. EPA 440/5-81-110. U.S. Environmental Protection Agency. Washington. DC., pp. 25–36.

Watt, W.D., Scott, D. and Ray, S. 1979. Acidification and other chemical changes in Halifax County lakes after 21 years. *Limnol. Oceanogr.* 24: 1154–1161.

Watt, W.D., Scott, D. and White, J.W. 1983. Evidence of acidification of some Nova Scotian rivers and its impact on Atlantic salmon (*Salmo salar*). *Can. J. Fish Aquat. Sci.* 40: 462–473.

Resource Management and Land Use Options in Mountain Watersheds: GIS Based Simulation Modelling

H. Schreier, S. Brown, W.A. Thompson and I. Vertinsky

ABSTRACT

Rapid identification and resolution of landuse conflicts in headwater regions provides the key for sustainable watershed management. GIS-based models were developed to document forestry, wildlife and recreation conflicts in the Tangier watershed in the North Columbia Mountains of British Columbia. A 120-year forest growth and harvesting plan was developed and with this model landuse conflicts between forest harvesting, caribou habitat and heli-skiing were identified, and possible options to resolve these conflicts were examined. A number of management scenarios with different resource use constraints were tested and a comparison was made between timber and non-timber values using the multiple accounts approach. The GIS-based simulation modelling is proving to be a powerful decision support tool.

Keywords: Resource management, landuse, simulation modelling, GIS, North Columbia mountains.

INTRODUCTION AND BACKGROUND

Geographic information systems (GIS) were combined with simulation modelling to evaluate conflicts between competing landuses. The North Columbia Mountains in British Columbia are currently experiencing growing conflicts between logging, wildlife, heli-skiing, snow mobiling, hunting and summer recreation. The area has a very active forest industry and timber shortages have recently been identified (B.C. Ministry of Forests, 1993). The wood processing capacity in the area exceeds the annual sustainable supply by 30% and this puts into question the long-term sustainability of the forest. The area is one of the most popular heli-ski and snow-mobiling destinations in North America, with a steadily increasing demand for such activities. At the same time many mammals use the area as their key habitat. This is particularly the case with the Woodland caribou, which reside in the adjacent national parks during the summer and use the surrounding areas for winter habitat.

A comprehensive geo-referenced GIS resource database was developed for the Tangier watershed which is located between Mount Revelstoke and Glacier National Parks in British Columbia (Fig. 1). The resource database consists of topographic information, forest inventory data, wildlife habitat information for caribou, climatic information and designated ski-runs for heli-skiing. The 28700 ha size watershed was divided into 850 management units based forest cover and topographic conditions. The forest growth was simulated over a 120-year period using local growth values for 10 different tree species. Since forestry has been the key resource user with the greatest potential impact, five different harvesting scenarios were examined. Each scenario has different constraints and emphasis. The results were compared using the multiple accounts framework, which included measures of timber, caribou habitat and winter recreation benefits. Heli-ski information was obtained from interviews with skiers and operators. Caribou habitat information and telemetry data of individual animal sightings were provided by wildlife experts with Parks Canada and the B.C. Ministry

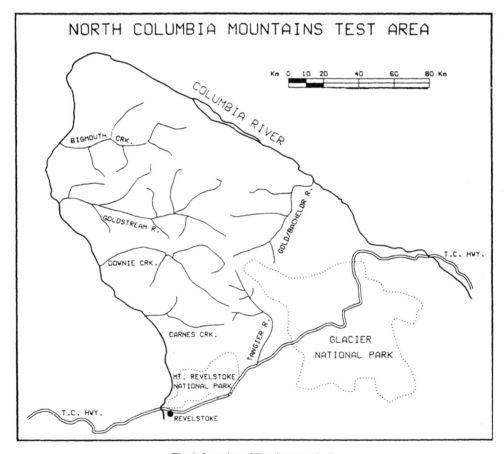

Fig. 1: Location of Tangier watershed.

of Forests, the forest inventory and yield information was made available from the B.C. Ministry of Forests.

The landuse and resource dynamics were modelled within the geo-referenced database, and the forest production and economics were determined over a 120-year rotation cycle. The two were linked to portray dynamic time snapshots of the resource status and conflicts. There are seasonal changes in the landuse for caribou and heli-skiing and the standing biomass for timber shifts according to growth rate and harvesting. All this can be captured and incorporated by linking the models with GIS and with the geo-referenced database.

The aims of the paper are to: 1. illustrate how the GIS-based model can be used to minimize landuse conflicts between forestry, caribou and heli-skiing; 2. determine how management constraint affects different uses; and 3. determine quantitative trade-off comparisons between uses with timber and non-timber values.

FOREST DYNAMICS

The forest inventory database was used to determine the current forest status. Forest stand attributes were related to stand age and included merchantable volume, dominant tree height and crown closure. The site quality and dominant species information was incorporated in the TIPSY stand growth model (Mitchell *et al.*, 1992) and the following five management scenarios were examined:

1. Green-up and adjacency constraints (G&A)
2. Even flow harvesting (Even)
3. Harvesting with caribou constraints (Caribou)
4. Old Growth conservation constraints (Old Growth)
5. Designation of conservation area within the watershed (Park).

Restricting the harvesting of any designated block of timber when any adjacent block is less than 20 years of age was the basic rule assigned to all five scenarios. The harvest scheduling program ATLAS (Nelson, 1993) was used as a framework to cutblock layout and to determine harvest schedules. In the first scenario (G&A) timber is harvested aggressively within the constraints described above. In the even flow scenario (Even) timber harvesting is maintained at an even annual rate over ± 2% of the timber volume over the entire 120-year rotation. With the caribou constraints scenario (Caribou) a minimum of 600 ha of high quality habitat is maintained at all times. In the Old Growth scenario 60% of the current old growth is maintained at all times. Finally, in the park conservation scenario 83% of the northern portion of the watershed is designated as park and only the remaining area is accessible for multiple use.

Information used to determine harvesting costs included road construction and maintenance costs (RC), tree to truck (TTC) and log hauling costs (HC). Log values (LV) were determined using average values for each species, site class and age class and converted into 1992 dollars based on the GDP price deflator. Finally, the net present value of commercial timber ($/m^3) was calculated using the following formula:

$$\text{Net present value} = LV - TTC - HC - RC/(A - V) - CA$$

where V= Merchantable timber (m³/ha), A= Area from which timber is removed (ha), and CA = Administrative cost ($/m³). Economic valuation of timber was made in real 1992 dollars assuming a 5% per annum discount rate and no change in real cost or price.

The results from the different scenarios were then compared using the multiple accounts method (Gunton *et al.*, 1991). In our case the trade-off between net present values for forestry and the amount of high quality caribou habitat retained by each management scenario were compared (Thompson *et al.*, 1993; Brown *et al.*, 1994).

CARIBOU HABITAT SUITABILITY ASSESSMENT

Caribou are highly migratory animals and they move during the winter according to environmental conditions and available food supplies. The approach used is based on species-habitat relationships as observed by local wildlife experts over the past 10 years (Simpson and Wood, 1987; Simpson and McLellan, 1990). Three following key periods were identified as critical habitat conditions over the winter period: Early winter (Nov–Jan 15), Late winter (Jan. 16–April 15) and spring (April 16–June 30). For each of these periods a set of topographic and vegetative factors were identified that reflect optimum, moderate and low habitat conditions. In early winter, when the snow is deep and soft, dense old growth forest at low elevation (< 1200 m) was considered the best habitat for caribou. In the late winter, when snow densification enables the animals to be more mobile higher elevation sites dominated by spruce and true fir sites with lower crown closure classes were considered optimum and during spring new food sources at low elevation riparian sites were considered best (for more details on the classification see Schreier *et al.*, 1993 and Brown *et al.*, 1993, 1994). Table 1 provides an example of the early winter habitat suitability classification.

For each of the three critical seasons a high, medium and low habitat suitability classification was produced using the GIS system. Using the GIS overlay technique a overall habitat suitability classification was produced by combining the three individual classes. The individual and combined classifications were then tested by superimposing caribou radio-tracking data from 35 animals. These animals have been monitored over the past 18 months on a bi-weekly basis and the positional data could readily be superimposed on the suitability classification within the GIS. This provided a unique opportunity to arrive at an independent assessment of the habitat suitability model.

HELI-SKI INFORMATION

All heli-ski operators require a licence to operate on crown land, and are required to identify individual runs and keep records of frequency of use during the winter. The locations of the licensed runs were digitized and incorporated into the GIS database and

Table 1: Criteria used for early winter caribou habitat suitability assessment

Variable	Suitability			
	High	Medium	Low	Nil
Elevation (m)	0–1199	1200–1499	1500–1799	≥1800
Slope (%)	≤ 29	30–39	40–69	≥ 70
Slope position	Lower	Mid valley	Upper	Crest
Dominant species	Spruce true fir Cedar hemlock	—	Douglas fir Lodge pine whitbark	Other
Height class	≤ 3	—	2	1
Crown closure class	6–10	—	4–5	0–3

the user information was linked to each run. Buffer zones of 500 m were created around each run and the total vertical metres per run were determined from the GIS.

The forestry, caribou and heli-ski information were then combined and analyzed in several comparative tests.

RESULTS

The forest production and harvesting evaluation based on the five management scenarios were first evaluated over the 120-year and the comparisons expressed in total harvest volume (m³) or net present value is provided in Figure 2. The best harvest volume and economically most favourable option is provided by the even flow scenario, while the park conservation option produced the least favourable option of forestry.

The amount of prime caribou habitat conserved over the 120-year period could also be examined for each scenario and each decade with the GIS system. The results from this comparison showed that the old growth retention was the best option while the green-up and adjacency scenario produced the least amount of caribou habitat over the entire time period.

If we use the multiple account system to compare the amount of prime caribou habitat retained by each management scenario with the net present value for forestry (Fig. 3), it is clearly evident that the even flow, caribou constraints and only growth scenarios produced the best combined options while the green-up and adjacency was the least favourable option for caribou habitat retention. In contrast, the proposed park was least profitable for forestry and did not retain as much habitat as the old growth, even flow and Caribou scenarios.

With the GIS it was possible to produce a combined multiple use conflict evaluation (Fig. 4). It contains spatial information on caribou habitat, actual caribou

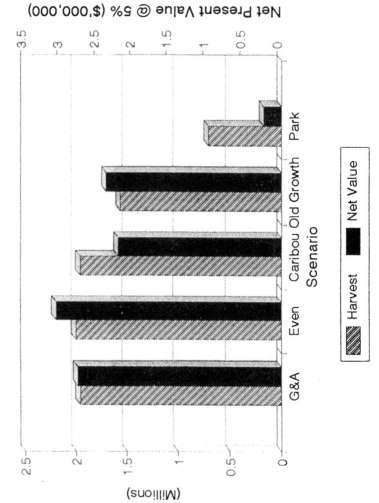

Fig. 2: Comparison management scenarios for timber production.

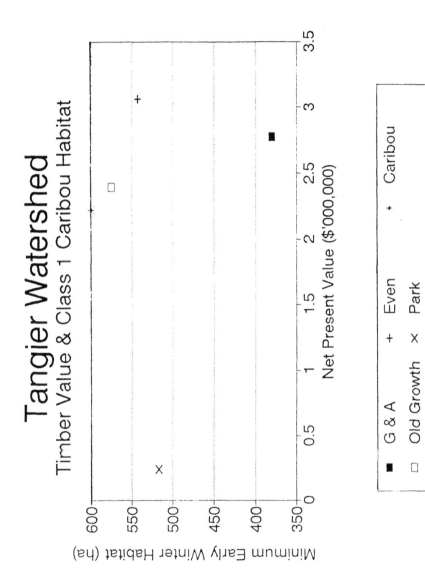

Fig. 3: Comparing net present value for forestry with high quality caribou habitat in multiple accounts framework.

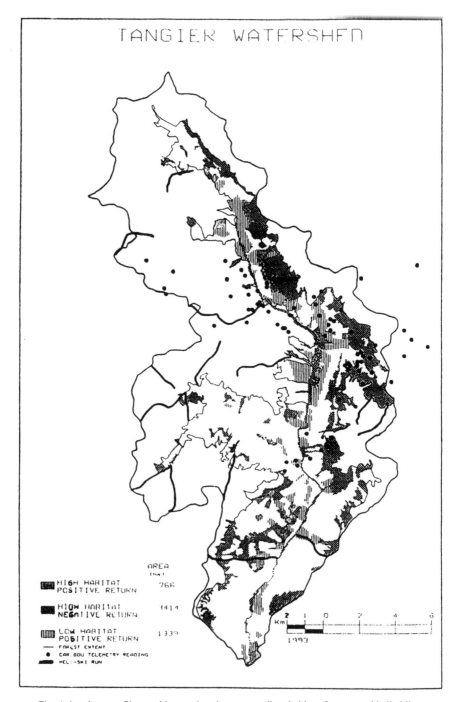

Fig. 4: Landuse conflicts and interactions between caribou habitat, forestry and heli-skiing.

locations, forests with positive and negative net present value, and designated heli-ski runs. Some 766 ha were identified as caribou/timber conflict areas where both caribou habitat and net present value for timber were high. Some 3414 ha were identified as high quality caribou habitat with low timber values and 1339 ha were shown to have high timber value but low caribou habitat suitability. These areas do not represent conflicts and can readily be designated to the respective uses. Finally, caribou and heli-ski conflicts can readily be shown by comparing the ski runs and the caribou locations as indicated by the black dots. At least one ski run in the NW portion of the watershed has caribou sightings within the ski area.

The advantage of the GIS based model is that multiple uses can be displayed in a combined manner or in a time sequence. For example, seasonal conflicts between caribou/heli-skiing can be shown for any season (early or late winter), and caribou/forestry conflicts can be displayed spatially for any time period over the 120-year rotation.

CONCLUSIONS

Combining GIS resource databases with simulation models provide us with a unique capability of assessing multi-use conflicts and management options in a watershed context.

1. The method is flexible and enables us to produce selective and integrated maps of landuses and landuse conflicts in a rapid manner.
2. The simulation capability enables planners and managers to develop management scenarios with different constraints that can readily be compared.
3. The multiple account analysis enables us to compare economic values for forestry with caribou habitat and other quantitative indicators of non-timber values such as old growth conservation or recreation activities.
4. The combined GIS linked model is a new and exciting decision support tool that may provide the managers and planners with a better set of options to make land use decisions in a more holistic and quantitative manner. These tools are well suited for resource management in headwater locations where landuse conflicts can have serious consequences within the headwater area as well as in the adjacent lowlands.

REFERENCES

B.C. Ministry of Forests. 1993. *Revelstoke TSA timber supply analysis.* Integrated Resource Branch, B.C. Ministry of Forests, Victoria, B.C., 60 pp.

Brown, S.J., Schreier, H., Woods, G., Thompson, W., and Vertinsky, I., 1993. A GIS approach to reslove resource conflicts in the trans-boundary area of Glacier and Revelstoke National Parks. In: *GIS-93 Itern. Symp. on Geographic Information Systems in Forestry, Environment & Natural Resource Management,* Forestry Canada, Feb. 15–18, Vancouver, pp. 221–226.

Brown, S., Schreier, H., Thompson, W.A., and Vertinsky, I. 1994. Linking multiple accounts with GIS as a decision support system to resolve forestry/wildlife conflicts. *Journ. Environmental Management* (in press).

Gunton, T., van Kooten, G.C., and Flynn, S. 1991. Multiple accounts analysis and evaluation of forest landuse conflicts. In: *Forest Resource Commission. Background Paper,* Vol. 2, 21 pp.

Mitchell, K.J., Grout, S.E. MacDonald, R.N. and Watmough, C.A. 1992. *User's Guide for TIPSY: A Table Interpolation Program for Stand Yields.* B.C. Ministry of Forests, Victoria.

Nelson, J.D. 1993. Spatial Analysis of Harvesting Guidelines in the Revelstoke Timber Supply Area. In: *GIS'93 Symposium: Eyes on the Future.* February 15–18, 1993, Vancouver B.C., pp. 203–208.

Schreier, H., Brown, S., Thompson, W.A., and Heaver, C. 1993. *Resolving Forestry/Wildlife Conflicts in the Tangier Watershed Using Simulation Models Linked with GIS Techniques. Final Report,* B.C. Hydro and Power Authority, pp. 58.

Simpson, K. and McLellan, B.N. 1990. *Wildlife Habitat Iventory and Management Planning in Mount Revelstoke and Glacier National Parks.* Canadian Parks Service, 95 pp.

Simpson, K. and Woods, G.P. 1987. *Movements and Habitats of Caribou in the Mountains of Southern British Columbia.* Wildlife Branch, B.C. Ministry of Environment and Parks, Nelson, B.C. Wildlife Bulletin #B-57., 36 pp.

Thompson, W.A., van Kooten, G.C., Vertinsky, I., Brown, S., and Schreier, H. 1993. A preliminary economic model for evaluation of forest management in a geo-referenced framework. In: *GIS-93, International Symposium on Geographic Information Systems in Forestry and Natural Resource Management,* Forestry Canada, Vol. 1, pp. 209–219.

Integration of Remote Sensing and Digital Terrain Models for Land Cover Classifications, Thematic Mapping and Regional Planning in Val Malenco (Italian Alps)

Maria Luisa Paracchini and Sten Folving

ABSTRACT

Val Malenco represents a typical, complex landscape of the North Italian Alps. It has been chosen as a test area for integrated applications of a standard DTM, digitized terrain map information, and both SPOT and Landsat TM data for land cover mapping and landuse classifications, in regional planning and modelling. Primarily the study aims at providing methods for low cost accurate mapping of surface cover types and at defining methods for monitoring the general environmental conditions in an alpine landscape.

Keywords: Remote sensing, digital terrain models, thematic mapping, Italian Alps.

INTRODUCTION

It is often difficult to apply remotely sensed data to land use and environmental mapping in high mountain areas, and yet, it is practically impossible to obtain needed information by any other means. There is a growing awareness of the need for careful, continuous management of the fragile eco-systems in mountain watersheds, together with a still growing need for a continuous supply of fresh water and hydro-electrical energy from these areas. This requires the development of applicable methods for using remotely sensed data together with more traditional map-data in landscape management and environmental protection.

The problem of floods, landslides etc. in mountains have always existed, it is part of the Alpine environment. The present change in landuse in European Alpine areas, in the short term is believed to have a very positive effect on hydrological conditions and on the frequencies of landslides. The abolishing of the smaller alpine farms means a larger surface covered by woods which of course is a very protective factor. On the other hand, this will lead to a lower degree of landscape diversity. In the future, Alpine farming in Europe probably will be supported, in one way or another, to maintain a rural population in these areas and to maintain the general landuse patterns.

For a couple of decades, the Environmental Mapping and Modelling group of the Joint Research Centre of the European Commission has been heavily engaged in research in image processing and the development of methods for extracting information on land cover from remotely sensed data. This research has, during the last few years, changed into research and development on the application of remote sensing for planning and management of the environment, especially in the marginal areas of Europe, which also often constitute the most interesting regions from an ecological and environmental point of view. Uplands and mountain regions are examples of such areas. At the same time they constitute important hydrological catchment areas.

In the frame of the European Collaborative Programme, a study on landscape cover type classification and mapping of Alpine regions for hydrological planning and management has been initialized. The main aim of this study is to provide a fast (semi-automatic) method for assessing the main characteristics of the surface cover in alpine watersheds.

Digital terrain models are used for geometric correction of the remotely sensed data—LANDSAT TM and SPOT data. Thematic map data is used for aiding the classification and for accuracy checking. The classification results, together with the thematic map information and terrain models, constitute elements of the Geographical Information System that is being built in order to carry out modelling studies—primarily on run-off and erosion.

This presentation will concentrate on the image processing part of the study: geometrical correction and classification. However, it should be noted that the modelling considerations determine the selection and development of methods.

DESCRIPTION OF THE STUDY AREA

The Mallero Valley is located in the north of Italy (Fig. 1) inside the chain of the Alps. It is one of the lateral valleys of Valtellina. The valley has an extension of approximately 320 km². It is characterized by steep slopes, especially in the upper parts near the Massifs of Bernina and Disgrazia, which combined with the heavy rate of rain in the area, cause serious slope instability. (In 1987 a great landslide (40×10^6 m³) occurred in northern Valtellina causing the death of 27 people and damming the course of the Adda River creating an impoundment of about 20×10^6 m³.)

The area is made up of intrusive and metamorphic rocks which are completely impervious. Underground drainage is only possible in fractured and clastic areas. The glaciers of Mount Disgrazia (3678 m) and Bernina (4050 m) form a regulating buffer on the precipitation when looked upon over a yearly— or longer—time span.

The Mallero Valley is situated in one of the most rainy areas in Italy. The number of rainy days is generally higher than 100 per year and the annual total precipitation often exceed 1000 mm—Fig. 2. The rainfall is concentrated mainly in the months of

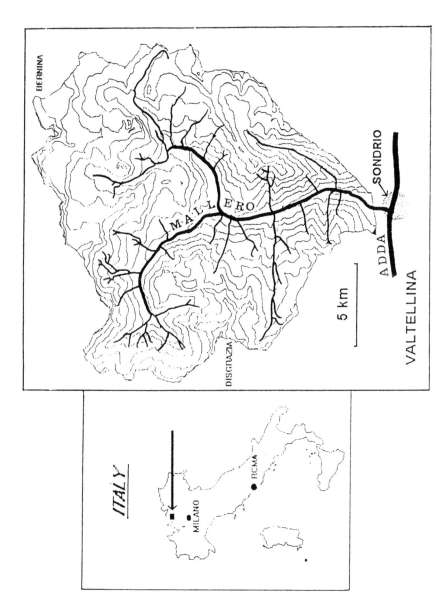

Fig. 1: The location of the Val Malenco test site in the Italian Alps. The topography and main drainage pattern is shown.

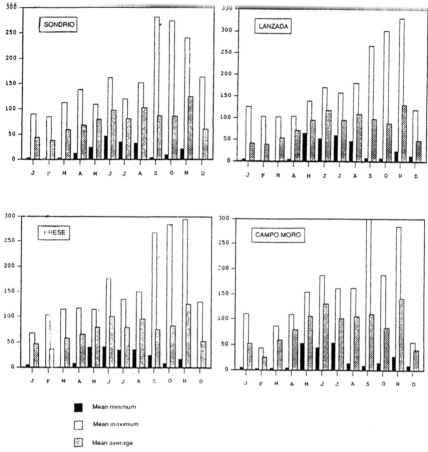

Fig. 2: Precipitation in the Val Malenco area. Mean data from the period 1958–1973.

June and November. The extremely heavy precipitation frequently causes floods, and this is exactly the reason why this area was chosen for testing the use of satellite based land cover mapping for hydrological modelling.

VEGETATION

The tree species which grow in the alpine area are distributed according to altitude and slope orientation (Fig. 3). In Val Malenco, the vegetation located in the belt between 250 and 900 m consists of mixed broadleaved forests composed primarily of: *Castanea sativa, Quercus* spp., *Fagus sylvaticus, Robinia pseudoacacia, Laburnum* spp. In the belt located between 900 and 1200 m, we find mixed broadleaved and coniferous forest (in which *Fagus sylvaticus, Acer pseudoplatanus and Abies alba* are predominant).

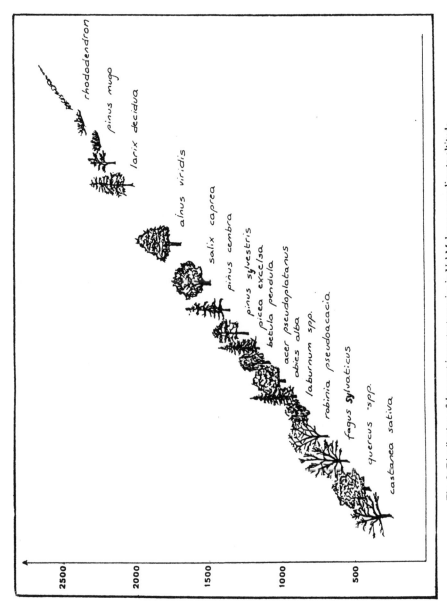

Fig. 3: Distribution of the most important species in Val Malenco according to altitude.

From 1200 upwards the broadleaves tend to thin down in number, leaving space for evergreens (*Picea excelsa, Pinus sylvestris, Pinus cembra*) and to the *Larix decidua*; the only broadleaved species that can be found up to 2000 m are the *Salix caprea* and *Alnus viridis*. Above such a height, only scattered *Larix decidua* is found (up to 2200 m) mixed also with *Pinus mugo* and rhododendrons. At 2450 m the ground remains covered by pastures, rocks and bare soil up to the limit of perennial snow.

The Val Malenco has a north-south orientation and therefore the eastern and western slopes are covered by the same type and density of vegetation, while in the lateral valleys, oriented in the east-west direction, the vegetation covering the slopes can often present very different characteristics. On the slopes which face north very thick forests are growing, while on those facing south primarily grassland and pastures are found. Here the woods are very rarely covering the slopes and it is not unusual to find isolated trees. This is caused by the differences in microclimate; but also by the human impact, primarily deforestation which causes extension of the pastures and cultivated areas.

SATELLITE IMAGES

Because of the heavy cloud cover in the area, it is often very difficult to find useful images for land cover classifications. Furthermore, the period suited for vegetation studies is restricted to the summer months due to the presence of snow on until late spring or early summer. It has not been possible to choose the images from the best periods for forestry applications, which would be the period of late May–early June, and—if multitemporal datasets would be required—August/September. The images which have been chosen for this study were recorded on 18/8/90 and 19/9/90.

The aim of this part of the project is to build a classification methodology to determine the areas covered by the land cover classes defined as useful for hydrogeological studies. Because we are looking for operational applications of the method concurrently with the development, it has been necessary to limit the pre-processing work. This means that atmospheric corrections have been avoided and that the classification does not fully take into consideration shadow effects. Both are very important factors in mountain regions. It is, however, indispensable to carry out geometric correction of the images chosen for the classification since the relief causes strong parallax effects. When proceeding to multitemporal images such geometrical correction becomes very important.

DATA CORRECTION

As we are dealing with a mountainous landscape, a digital terrain model becomes necessary in order to perform geometrical corrections of the imagery and in order to calculate correct areas, slopes and aspects and to be able to make corrections for illumination/shadow effects.

The digital elevation model was established from digital contour lines of the area. The data was bought from the Italian Military Geographical Institute. The contour lines were digitized at scale 1:100000 with an equidistance of 50 m. Several map sheets had to be merged. In glaciated areas, and in a few places not covered by the purchased data, the elevation data were captured by manually digitizing 1:25000 maps. All digitization and the creation of the model was done using ARC/INFO. After the model was built it was transferred into raster format in order to facilitate the necessary further computations using standard image processing software.

The elevation model has been used for creating the normal terrain surface, slope, slope length and exposure, but it is also used as an expert reference for classifications in deciding whether given vegetation classes can be accepted above a certain height, it is used to refine the classification.

As shown on Figure 4, the DTM was indispensable for correcting the satellite image. In an area like Val Malenco, the parallax error easily reaches 300 m. The satellite data—and any other data to be built into the GIS datalayers has to be geo-referenced. The correction for parallax has been performed with specific software developed at the Joint Research Centre (Hill *et al.*, 1989). The procedure consists of three main steps: a preliminary, rough transformation is performed using a matrix of coefficients derived from ground control points; then the DTM is integrated for a per pixel calculation of the distance between the position of the pixel on the image and their correct geographical position; and, finally the effective transformation of the image data and resampling of the radiance data is performed. After the correction, the image still contained a maximum error of approximately two pixels (60 m) in the higher summits which has been accepted for this study, as the main interest concerns the central and lower parts of the mountains.

Classification experiments showed that only bands 2, 4, 5 and 7 needed to be included. The first four components, from a principal component analysis (PCA)

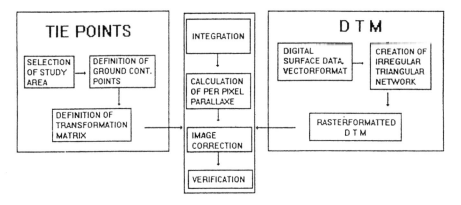

Fig. 4: The integration of the digital terrain model and the satellite imagery. Tie points are used for a preliminary matching, the DTM is used on a per pixel basis.

transformation of the total TM data set, gave almost the same result. It also turned out that a simple histogram equalization improved the result.

The classification used is a hierarchical procedure based on spectral mixture modelling (Adams and Smith, 1986; Smith *et al.*, 1990). This technique has been considered particularly useful for the type of application aimed at in the present study because the classification should include both vegetation type and density.

The linear mixture modelling is a method that permits calculation of the percentage of the components whose signal build up the radiance of the pixel. It is assumed that the single pixel reflectance is a linear combination of the reflectance of its single components. The vector representing the radiance of the pixel in the different spectral bands corresponds to the weighted mean of the radiance of all the cover types existing inside that pixel. The method makes it possible to obtain a new vector representing the percentages of the components corresponding to those cover types:

$$x_i = E_{i(i=1,n)} f_i M_{ij} + e_i$$

where n is the number of bands, x_i is the signal value in band i, f_i is the percentage of pixel area occupied by cover j, the M_{ij} coefficients are called "endmembers" and represent the signal coming from cover j pure pixels, and e_i is a noise term.

The method gives an estimate of f_i using M_{ij} values as input. There are two important constraints. The first is that the pixel is assumed to be composed only by the cover types we are taking into consideration. The second is that the value of the pixel must lie between 0 and 1.

Four endmembers were extracted to perform this part of the classification: one for bare soil, one for grassland and two for forest (broadleaved and coniferous respectively). This division of forest into two sub-classes was necessary to distinguish meadows from broadleaved in full sun. At a later stage the result of this distinction was regrouped to obtain only one forest component. The endmembers were chosen directly from the images according to "ground truth" data.

The classes chosen in the classification procedure are: snow, water, bare soil or rocks, forest and grassland. For the last three classes, intermediate classes were calculated according to variable vegetation density. In practice each pixel was considered to be composed of a variable percentage of two classes x and y (it was chosen to ignore the presence of pixels made up of more than two elements at the same time), whose distribution has been subdivided according to the following (see also Fig. 5):

4/4—pixel containing x > 75% and y < 25%
3/1—pixel containing 50% < x < 75% and 25% < y < 50%
1/3—pixel containing 25% < x < 50% and 50% < y < 75%

The mixed class bare soil/grassland was treated in a special way. Because it is possible to classify with high accuracy the unvegetated areas above a certain height, a distinction could be made between pixels made up of 100% bare rocks and pixels containing at least 3/4 of bare soil. For this purpose, the first three principal components

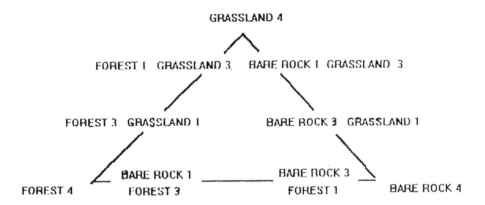

Fig. 5: Diagram illustrating the classification concept used in the study; see text for explanation.

were used since, in the composition, the difference between vegetation and bare soil is extremely accentuated.

The classification procedure is a sort of hierarchical analysis in which one or more classes are extracted at each step. A mask is created and the following classification is carried out on the image from which already classified areas have been masked. The last classes to be extracted were the mixture of forest and grassland. These are complex to define and classify since in full sun, shaded or slightly shaded grassland and pastures have reflectance values very similar to those of broadleaved forest.

The classification included the areas either fully covered by vegetation or mixed with some bare surface types, (Fig. 6). The areal distribution of the surface cover types according to the described classification is shown in Table 1.

PROVISIONAL APPLICATION

It should be emphasized that the application results presented in the following paragraphs are provisional. The U.S. Soil Conservation Service "Curve Number" (CN) model has been used (USDA, 1972). This is a fairly simple model, it requires information only on soil type and on land use. The model was developed for smaller watersheds in agricultural areas. Therefore it is most probably far too simple for Val Malenco, but will serve as a preliminary tool.

The total run-off in mm, Q, is given by:

$$Q = \frac{(P - Ia)^2}{(P - Ia) + S}$$

Table 1: The main classes used in the study and their areal cover

Class	Hectares
Forest 3—grassland 1	3560.4
Forest 4/4	5981.4
Grassland 3—forest 1	402.6
Grassland 4/4	3370.5
Forest 3—bare soil 1	501.0
Forest 1—bare soil 3	1343.0
Grassland 1—bare soil 3	1279.6
Grassland 3—bare soil 1	1071.2
Bare soil 4/4, shadow, rock and urban	11554.1
Water	148.7
Snow	3073.1

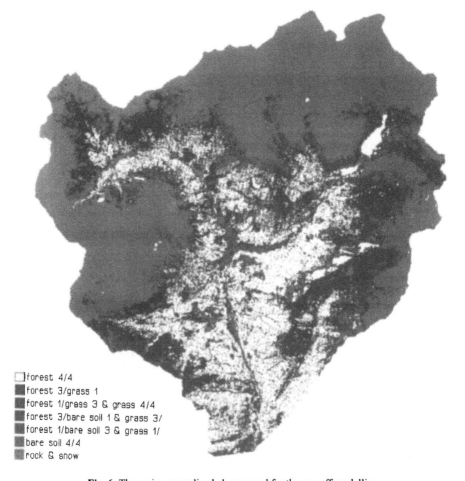

forest 4/4
forest 3/grass 1
forest 1/grass 3 & grass 4/4
forest 3/bare soil 1 & grass 3/
forest 1/bare soil 3 & grass 1/
bare soil 4/4
rock & snow

Fig. 6: The main, generalized classes used for the run-off modelling.

where P is precipitation, Ia is a initial interception and S (mm) is a parameter defined as the sum of Ia and the maximum potential retention of the watershed. The initial interception, which is empirically determined, is defined as:

$$Ia = 0.2 \, S$$

which leads to:

$$Q = \frac{(P - 0.2 \, S)^2}{(P + 0.8 \, S)}$$

S is standardized to variate between 0 and 100 by introducing the CN parameter:

$$CN = \frac{25400}{S + 254}$$

CN is dimensionless and describes the combined class and land cover type. The larger the variable CN the larger the potential run-off.

The results from the land cover classification have been re-coded into retention classes. A rather pragmatic approach was used. Bare rocks and ice were given a retention capacity of 0% and the most densely covered forest areas was given a capacity of 100%, which will only be valid for normal rain intensities during short periods. The re-coding into an ordinal scale is rather experimental.

The new classes are listed in Table 2. The areas in shadow, which were not correctly classified in the original classification, have been treated as areas without information and are not included. The water covered areas have been neglected in this regrouping as a priori they have a CN of 100. The permeability is one of the more important parameters to be included in hydrological modelling. The classification of the soils into hydrological classes as proposed by U.S. SCS is based on infiltration capacity and has been done for all the soils in the USA. This cannot be done in Val Malenco as no soil map exists. However, a permeability map in scale 1:100000 does exist from this watershed and has been manually digitized. It describes three permeability classes (from high to medium, from medium to reduced, and, from reduced to very reduced) of the solid rocky surface and, correspondingly, three classes for loose deposits. These six classes are re-grouped into four classes of infiltration which is the number of classes used by the CN method. Group A, consisting of loose deposits with

Table 2: The reclassification of the land cover into retention classes

1. 100% retention	=	forest 4/4
2. 85% retention	=	forest 3—grassland 1
3. 60% retention	=	forest 1—grassland 3 & 4/4
4. 40% retention	=	grassland 3—bare soil 1 & forest 3—bare soil 1
5. 20% retention	=	forest 1—bare soil 3 & grassland 1—bare soil 3
6. 10% retention	=	bare soil 4/4
7. 5% retention	=	snow and rocks

high to medium permeability, is given an infiltration capacity of 100%. Group B, which consists of loose deposits of medium to reduced permeability, is given an infiltration capacity of 75%. Group C, contains loose deposits of reduced to nonexisting permeability and solid surfaces with high to medium permeability, has been given the value of 55%. Finally, group D, which has been given the capacity of 10% consists of solid surfaces with medium to nonexisting permeability. All the infiltration values should be regarded as indicative and, like the retention classes, treated as qualitative data on an ordinal scale.

The two classifications, or re-groupings, now represent two strata in raster format, describing some physical geographical parameters in Val Malenco. A simple model applied to these two strata results in a provisional "hydrological surface cover complex" for each pixel in the valley, Table 3. The values in the pixel represents run-off indices, the corresponding map is shown in Fig. 7.

Table 3: The run-off indices; see text for explanation

		Infiltration			
		A	B	C	D
Retention	%	100	75	55	10
forest 4/4	100	200	175	155	110
forest 3—grassland 1	85	185	160	140	95
forest 1—grassland 3 & grass 4/4	60	160	135	115	70
grass 3—bare soil 1 & forest 3—bare soil 1	40	140	115	95	50
forest 1—bare soil 3 & grass 1—bare soil 3	20	120	95	75	30
bare soil 4/4	10	110	85	65	20
snow & rock	5	105	80	60	15

This new classification simply states that the higher the value of the pixel, the smaller is the run-off from that pixel.

CONCLUDING REMARKS

The study has not been finished yet. This spring (1994), the results were given to ISMES, a private Italian company with a branch specialized in environmental management. ISMES is incharge of monitoring the whole Valtellina watershed. They will implement the classification results and, at a later stage, the classification methodology in their operational and experimental watershed management strategy. As a first step, it has been planned to include the land cover mapping into the various models used for predicting the run-off in order to evaluate if an improvement can be provided. The next step will be to transfer and test the methodology in other sub-watersheds.

Fig. 7: Map of the run-off indices in the Val Mallero watershed. The white areas represents the glaciers; dark areas have high run-off indices and light gray represents low run-off indices.

ACKNOWLEDGEMENT

The authors express their gratitude to their colleagues in the Environmental Mapping and Modelling Unit at the Joint Research Centre for assistance and support.

REFERENCES

Adams, J.B. and Smith, M.O. 1986. Spectral mixture modelling: A new analysis of rock and soil types at the Viking Lander 1 site. *Journal of Geophysical Research*. Vol. 91, no. B. 8.

Hill, J., Kohl, H., and Mégier, J. 1989. The use of the digital terrain model for landuse mapping in the Ardeche area of France. *Proceedings of the 9th. EARSeL Symp. Espoo, Finland*, 27 June–1 July.

Smith, M.O., Ustin, S.L., Adams, J.B. and Gillespie, L.A. 1990. Vegetation in deserts: I, A regional measure of abundance from multispectral images. *Remote Sensing of Environment*. Vol. 31.

US Department of Agriculture, Soil Conservation Service. 1972. *National Engineering Handbook*, Section 4, Hydrology. US Governmental Printing Office. Washington, USA.

Problems of Headwater Management in the Central Himalaya

Martin J. Haigh

ABSTRACT

Political controversy ranks among the most fundamental problems of headwater management. The manager faces pressure from many different groups, each with their own world views, their own supporting evidence, and their own policies. This paper examines the ideas and beliefs of some of the rival factions who seek to influence headwater management policy in the Central Himalaya.

Keywords: Himalaya, watershed management, policy, environmental controversy.

INTRODUCTION

Successful land management is difficult to achieve even when the manager has access to perfect data. Unfortunately, this situation is rarely encountered outside textbooks. Real decisions are formulated in a haze of uncertain and imperfect information as a compromise between perceived necessity and political/economic expediency. They come to depend upon, not so much objective reality, but the world view of the decision maker. Problems arise when rival environmental managers with different beliefs, based on different premises and different factual evidence, approach the same problem.

The Central Himalaya provides a useful case study. Here, the coexistence of rival world views, and the absence of any universally agreed corpus of hard factual data, are the source of many of the headaches in headwater management (cf. Bruijnzeel and Brenner, 1989; Bruijnzeel, 1990). The greatest division separates the environmental management establishment from the NGO environmentalist both at the national and international level (cf. Dogra, 1992; Haigh, 1989; Metzner, 1993). However, a huge range of divergent thinking also exists within each camp. The disagreements throw into question what is known and what is not known about the state of the environment, what are the problems—if they exist, what are the causes and, inevitably, what should be the solutions. So, although, all parties in the Himalaya debate espouse a single objective: the regeneration/preservation of a sustainable, productive, high quality, environment in the Himalayan mountains, there is less agreement about how this may be obtained.

TWO VIEWPOINTS ON FOREST CONSERVATION

Poldane (1987) compares the ideas/ideologies of India's official Forest Department with those of their most vocal, non-government critics, the Chipko Movement. Chipko is a popular environmentalist group, founded on Gandhian principles, devoted to forest conservation and rural uplift according to the Sarvodaya model (Hegde, 1988; Misra and Tripathi, 1978; Doctor, 1965). The movement combines political activity with practical development projects (Haigh, 1988). As Shah (1982:12) comments: there is a divergence between the rationality of the hill farmers and those of the national agencies. Poldane summarises the results of a programme of interviews as a table. A modified version follows as Table 1.

Table 1: Comparing the world views of the Official Forest Department with that of the Non-government Environmentalist Group: "Chipko" (after Poldane, 1987)

Official Forest Department	Chipko Movement Environmentalists
Broadly Socialist—planned, scientific management is the route to meeting the basic needs of the people and providing environmental protection	Broadly Gandhian—development involves the harmonisation of human society within nature and meeting basic needs through the elimination of non-essential wants (cf. Doctor, 1967)
Economic growth is essential	Economic growth is irrelevant
The Himalaya and the Plains are part of the single national unit. India's needs over-ride local interests	The social welfare and uplift of the local community is the prime consideration
The forests are a national asset and they are managed for the benefit of the whole nation	The forests are a village resource but they are exploited for the needs of big business and the cities
Environmental stress is growing because of local population growth and the "tragedy of the commons" (Shah, 1982:13)	Environmental stress has been caused by the expropriation of traditional village lands for commercial forestry. The villages have been forced to survive on a land base which is too small
Forest degradation is due to the short-sightedness of local village communities	Forest degradation is caused by external national/commercial forces and a policy of revenue maximisation (cf. Ahmed, 1985)
Government is necessary to protect the forests and to ensure they are properly and scientifically managed	The official forest operation is inefficient, ineffective, and locally corrupt. Current practices cause serious environmental damage
Current development strategies of commercial and industrial development are adequate	Current development trends encourage the disruption and exploitation of both environment and society
Improved scientific management practices, better forest policing, greater investment of national resources	Ban on commercial timber production, community-oriented afforestation, development of alternatives to forest produce (e.g. developing small-scale hydroelectric capacity to reduce demand for firewood)

THE HIMALAYAN CRISIS: AN INTERNATIONAL DEBATE

Two further viewpoints may be discovered in the writings which originate outside the Himalayan region. Western environmentalists, representing often political, but essentially non-government/community-oriented activism, argue that the Himalaya is afflicted by an accelerating spiral of environmental decline (Reiger, 1976; Eckholm, 1976; Ehrich, 1989; Myers, 1984, 1986). Opposing this world view are a group of academics linked to Government organisations and the established structures of international development aid. This group argues that while there is a serious environmental problem in the Himalaya, claims about its scale and extent are much exaggerated. There is no "super crisis", just a problem which merits careful analysis and skilful treatment. The International Development Establishment (IDE) accuses the environmentalists of sensationalism and reflects upon a "haze of uncertainty" surrounding many of the "facts" about the Himalaya. Emblematic is the variation between the highest and lowest estimates for per capita fuel consumption in Nepal—they vary by a factor of 26 (Ives, 1987, 1988; Thompson and Warburton, 1985; cf. Bartwal and Bhatt, 1987). For their part, the environmentalists accuse the IDE scientists of 'complacency' and 'fudging myopia'. The two world views are contrasted as Table 2.

FIELD OBSERVATIONS BY THE HIMALAYAN ACADEMIC COMMUNITY

At present, the main sources of field data on the state of the environment in the Himalaya are the scientists of the universities and the research institutes of the Himalayan nations, especially India. This local research is largely ignored by international science (Haigh, 1988). IDE leaders, Ives and Messerli (1989), include just 37 publications from India and Pakistan amongst the 325 sources cited in their book, "*The Himalayan Dilemma*". The research of the Himalaya's own academics is also neglected by many of the western environmentalists who tend to cite their S. Asian counterparts (i.e. Agarwal and Narain, 1985; Agarwal *et al.*, 1982; Bahuguna, 1985; Bhatt, 1989; Himalaya Seva Sangh, 1977–1995). However, these neglected works are produced by scholars who live in the hills, and who, in many instances, were born and raised in Himalayan communities. Their scientific inference is supported by long personal experience and tested by day-to-day local contact.

By and large, the Himalayan academics support the western environmentalists' models of the crisis in the Himalayan headwaters (Rai *et al.*, 1989; cf. Haigh *et al.*, 1990). The Chipko movement is an important local force. This movement grew up because it gave a voice to a community with a problem. One of its most significant features is that its main activists are women (Shiva, 1986). In Central Himalayan society, men hold the assigned power but the bulk of agricultural production and the collection of domestic fuel and fodder is accomplished by women. Ahmed (1985:30) suggests that 97% of the women are involved in cultivation as compared to 12% of the

Table 2: Comparison of the beliefs of the environmentalist and IDE establishment camps on the state of the Himalayan environment

Western Environmentalist	International Development Establishment
The felling of the forests of the Himalaya is one of the world's greatest ecological disasters ... (Myers, 1984:3)	A super crisis is not imminent in the Himalaya, the scene is set rather for the deepening of a series of common problems ... (Ives and Messerli, 1989:237)
The Himalayan environment is trapped in a spiral of geoecological collapse which involves massive deforestation, accelerated erosion, increased landsliding, flooding on the plains and the drying up of streams in the hills... (e.g. Eckholm, 1976:75-84)	The causal relationships between population growth, deforestation, loss of agricultural land and downstream effects ... most are founded upon latter-day myth or falsely based intuition, or are not supported by ... reliable data (Ives and Messerli, 1989:9; Ramsay 1987)
The cause of the crisis is commercial and colonial exploitation which has forced local agroecosystems into instability ... and it is due to "population pressure" (Myers, 1984; Ashish, 1979)	The crisis is something to do with the growing pressure on resources, given the available technology, and pervading institutional and political situation (Ives and Messerli, 1989:208)
Local people are the most reliable source of information and of remedies	Western experts are the most reliable source of information and remedies
Community-lead activities are the best solution. Each village should be encouraged and assisted to develop its own lands for its own uplift, and to rediscover a harmonious balance with nature. Change must come from the people (Haigh, 1984:198)	Progress depends upon macro-scale watershed management, reduction of neutralisation of border conflicts, massive realignment of access to natural resources ... and effective peoples participation (Ives and Messerli, 1989:237)
Objective: ecodevelopment	Objective: economic development

men—the only task which is exclusively male is ploughing (Singh and Sharma, 1987:523). In general, women are only mobilised in times of extreme challenge (Ahmed, 1985:64). Few, in the male-dominated academic community contradict their message (cf. Chalise, 1986). Several echo the fears of the western environmentalists (e.g. Valdiya, 1985; Moddie, 1985; Negi, 1983).

Naturally, the writings of the Himalayan academic community contradict those of the International Development Establishment (IDE). One key area is the issue of deforestation. IDE writers do not like this term. Hamilton and Pearce (1986:75) grumble that the word "is used so ambiguously that it is virtually meaningless as a description of landuse change"... In fact, this is also true of the word "forest". Much time has been spent in definition of this term in order to map changes in forest area. India's official statistics suggest that 67% of the land area of hills, the Uttar Pradesh (UP) Hills, are forest. Unfortunately, much official "forest" is entirely devoid of trees. Gupta (1979) used satellite imagery to show that just 37.5% of the area actually supports forest trees. Tiwari *et al.* (1986) estimate 29% forest, but note that forest with a crown canopy of 60% accounted for only 4.4% of the land. Tiwari and Singh (1987) report a rate of agricultural extension into forest of 1.5%/year from their study of 200,000 ha of Kumaun. Significantly, even IDE researchers support this notion. Gilmour (1988:343) purports to "challenge the orthodoxy of widespread deforestation in the hills of Nepal". However, his case study data show a 1.3%/year loss of natural forest, somewhat offset by Government afforestation and commercial plantings. It's worth noting that private land is seen as a key to re-establishing forest in Swat and Dir, Pakistan (Haigh, 1991).

Hamilton and Pearce (1986) are also upset when the term "deforestation" describes commercial timber harvesting and protest that "generalizations" about the effects of harvesting in forests are difficult to make because of the many variables in precipitation patterns, topography, methods of product extraction and ... other factors (Hamilton and Pearce, 1986:92). However, field results from both India and Pakistan suggest that forest degradation reliably leads to reduced infiltration, more surface runoff, and higher flood peaks in affected rivers (cf. Haigh *et al.*, 1990; Sastry and Dhruva Narayana, 1986; Subba Rao *et al.*, 1985; Mathur and Sajwan, 1978; Masrur and Hanif, 1972; Agnihotri *et al.*, 1986; Pandey *et al.*, 1983; Rawat and Rawat, 1994). It also suggests that once soil is lost, reforestation may not reverse all of these changes for many years.

Any increased flow of flood waters in the Himalaya go to the plains but its importance to the Gangetic system is not fully established. However, the IDE (Ives *et al.*, 1989) are particularly adamant that deforestation in the Himalaya is not the cause of flooding in Bangladesh. Those who work in Bangladesh do not quite share that certainty, though they are convinced that there may be many factors contributing to the increase in that nation's flood problems. Kattelmann (1987) recognises C.K. Sharma as Nepal's leading authority on river hydrology. Sharma (1981) writes that: "heavy rainfall, coupled with deforestation . . . reduce the low flow of rivers and increase the high flow during the flood seasons. Most of the rivers of the Terai (the plainward margin of the Himalaya) previously contained water during the lean season, but are now found to be dry during winter". Not surprisingly, the Government of India's 1980 Commission on Floods recognised deforestation as one of the causes of increasing flooding on the plains of India.

The IDE's Ramsay (1987: 246-7) writes: "Locally, forest degradation and the conversion of forest ... has led to increased sediment mobilisation ... [but] such ... represents only a small portion of the total sediment contribution ...." Valdiya and Bartarya (1989:485) point out that a "greater than 13% reduction of protective cover on vulnerable hillslopes in 22 years, coupled with the excavation for roads and canals, has greatly accelerated erosion in the Gaula Catchment, Kumaun Himalaya. Their study of 548 sources of debris generation found most (5.57 per km^2) associated with roadways, followed by river beds (4.52 per km^2), and recently devegetated barren land (4.37 per km^2). By contrast, forests and agricultural terraces showed the fewest number of sediment sources (0.6 per km^2) (Bartarya, 1988, 1991; Valdiya and Bartarya,

1989:490). The author's own studies in the Arni Gadh catchment near Mussoorie found that sediment yields from slopes vegetated with scrub were 5–7 times greater than those which retained forest and suggested that partial (45%) "development" had increased the sediment load of the channel by about 7% (Haigh, 1982, 1979).

According to the Uttarakhand Seva Nidhi (1986:6), an NGO: "Environmental degradation has resulted in 55% of hill springs drying up in the past 20 years. The remainder have declining discharges". The IDE's Hamilton (1987: 258), under a subhead: "Myths and Misunderstanding" writes: "Cutting of forests ... has usually resulted in the free-water table moving closer to the surface, because less water is lost to evapotranspiration... Yet there is a popular belief that forest removal causes groundwater levels to lower, affecting wells and springs adversely" (cf. Hamilton, 1985). Valdiya and Bartarya (1989:417) report: "Deforestation...has affected the hydrologic regime of the Gaula catchment ... Kumaun ... This is manifest in the drying up of springs and diminished discharges in more that 40% of the villages of the Gaula catchment. The extent of decrease of discharge ranges from 25–75%" (see also Joshi et al., 1983; Dewan, 1990; Pant and Rawat, 1989; Valdiya, 1985a, b). The IDE's Thompson and Warburton (1985) write: "there are two ways of approaching the environmental crisis in the Himalaya. We can ask 'what are the facts' and we can ask 'what would you like the facts to be?' ".

Meanwhile, the Himalayan academics take issue against three rival camps on one important point. The IDE, the Gandhian environmentalists of Chipko, and many of the latter-day Western Environmentalists agree that the peasant farmers of the Himalaya have little responsibility for the current environmental problems of the region. These workers agree that the cause of the problems in the Himalayan headwaters lies outside the region—they are: legacies of colonial exploitation, mismanagement by official agencies, commercial pressures originating outside the region, the construction of roads and tourism/immigration (cf. Dogra, 1992, 1983). Silas et al. (1987) order local reasons for deforestation thus: commercial timber cutting, forest fires, overgrazing/lopping/extraction of resin/medicinal plants, agricultural extension, landslides/erosion—often due to road construction. The Himalayan academic community and the official environmental managers agree that a fundamental imbalance in the agroecosystems of the Himalaya is among the causes of accelerating environmental degradation.

The nature of this imbalance is described by a team which has investigated the energy balance of the Himalayan agroecosystem at the village level (Singh et al., 1984; Pandey and Singh, 1984, 1985) (cf. Gupta et al., 1982; Shah, 1989; Patnaik and Ramakrishnan, 1989). This study attempted to balance energy production by the agricultural system with energy consumption in three Kumaun villages. The results are most easily interpreted in terms of what is required for, and what is produced by, the cultivation of a single hectare of agricultural land.

Production requires several inputs. There include: seed, human labour, animal labour and manure. In the hills, the animals supply just over half of the total energy needed for agricultural production (Singh and Naik, 1987). The animals require fodder, only 10% of which can be supplied from the agricultural fields. The rest must be supplied from adjacent land. Singh *et al.* (1984) calculate that it would require the productivity of 18 ha of well-stocked forest or 30 ha of grassland/scrub to provide the animals sufficient fodder from free browsing.

However, each hectare must support 17.6 humans, who need timber to build houses and tools, and who need more fuel/firewood than is produced by the agricultural fields. Gupta *et al.* (1982), working at Bhaintan, Garhwal, found that 86% of the biomass extracted from local forest went to fodder/feed, 1% to timber etc, and 13% to fuel. Bartwal and Bhatt (1987) estimate that, in Garhwal, per capita fuelwood consumption is about 0.6 t/person/year, a ninth of the productivity of well managed forest. Singh *et al.* (1984) estimate that if the farmer were to have just 5 ha of well stocked forest, harvested at 30% of its annual production, there would be enough to supply all the cut-fodder, fuelwood, and timber to support production on 1 hectare.

Unfortunately, in the Central Himalaya, the ratio of agricultural to forest land is not 1:18 or even 1:5, but just 1: 1.7. Many workers have reported that forest lands are overgrazed, carrying perhaps a third more livestock than is consistent with their preservation (cf. Sharma, 1981; Hopkins, 1983). In Gorkha District, Nepal, Fox (1982:19) found grazing levels 3 to 9 times higher than those necessary for replenishment. Tiwari *et al.* (1987) note the damage caused by burning forest and scrub to improve grazing. The result is severe land degradation. Meanwhile, high agricultural terraces are abandoned because there is too little manure to preserve fertility.

Still, the agricultural output from these lands is not enough to support the people for the whole year. The result is that families have to export their sons to money earning enterprises outside the village, perhaps on the plains, in the Armed Forces. Gupta *et al.* (1982: 22) report from Bhaintan, Garhwal, that off-farm income is 40% of the cash value of the total farm product, and that out-migration amounts to one person in each two families (1:11.6). Rawat (1983) suggests that as much as 66% of the male population of the UP hills may be involved in economic migration (cf. Shah, 1986: 67% in Village Chausali). In Bironkhal Block, Pauri-Garhwal District, remittances drawn through the local post office accounted for 48% of village income, three times as much as agriculture. Rawat (1983) adds that 75% of the subsistence requirement of these villages was purchased. Ashish (1979) has named the result a "Money-order economy" because the survival of the village economy depends upon the funds sent home by those employed elsewhere.

SOLUTIONS?

Local environmentalists, like Ashish, argue that solving this problem requires a re-orientation of the local subsistence economy to cash-crop agriculture and that, as a first step, the Government should locate a fully stocked grain storehouse in each village. Chipko solutions include a ban on the export of timber, energy substitution through the provision of small-scale hydroelectric projects, active community afforestation of degraded village lands (Bandyopadhyay and Shiva, 1987), and major social reform

(Bahuguna, 1985). Community forestry, not commercial forestry, is the route to reestablishing ecological harmony in the hills (Shiva, 1987).

Western environmentalists tend to offer similar remedies (cf. Haigh, 1984). Nautiyal and Barbor (1985) argue that there is a need to curb population growth and transform the traditional land management system. Plantations must be established immediately to provide fuel and fodder for basic needs and to protect the land from further degradation. However, since these plantations must be established upon land which is already intensively grazed, free grazing must be replaced by stall-feeding with hand-cut fodder.

The Himalayan academic community broadly supports these analyses (Bartwal, 1987; Chaturvedi, 1985). Agroforestry, social forestry, and horticulture must be encouraged. The Hill's problems may also, perhaps, be alleviated by economic development, notably an expansion of tourism (Pant, 1983) and hitech industries (Prasad, 1983). This worries some who fear increased immigration into the hills and its impacts on an already over-stretched environment (Singh and Kaur, 1986).

The official viewpoint, at least in the Indian Himalaya, is enshrined in the Five Year Plans. These envisage massive electrification, which would reduce the dependence on fuelwood. The power would come from large hydel schemes like the controversial new Tehri High dam—a project universally disliked by environmentalists. The plans emphasise rural uplift through improved housing, sanitation, and water supply. They also envisage an increase in the road network. Road construction is named another major source of disruption in the Hills (Haigh, 1984; Ives and Messerli, 1989: 120) and roads are heavily criticised for facilitating illegal tree theft (Bahuguna, 1985). The Uttarakhand Seva Nidhi (1986: 9) a local community service NGO, however, agrees that roads are vitally necessary.

Government practitioners in the Himalaya worry about the Government custom of setting up offices with insufficient resources for the work required. They resent political interference which damages their long-term strategy and political pressure for short-term economic exploitation. They support the need for social forestry, overseen by trained foresters, supported by the provision of tree seedlings and nurseries. They agree that land reform is essential to the success of such projects. Common land is the bane of social forestry projects (Shah, 1982: 22; cf. Ashish, 1979). Greater involvement in the local community and improved provision of community assistance is also advised (Shah, 1982: 25), although the problems of working across the social divide of gender are less recognised.

The IDE recognises many of these issues. However, their remedies reflect a supranational perspective. They argue for international working groups to be set up, for the establishment of Peace Parks in the border areas between Himalayan nations, for the establishment of documentation centres/training institutes, for the creation of action plans in conjunction with major international agencies, and for a renewed emphasis on the collection of reliable research data (Ives *et al.*, 1989; Ives and Messerli, 1988: 270).

CONCLUSION

The morass of this controversy helps conceal the real situation in the Himalaya. The hard facts are contested. Case studies are discounted as biased. Overviews are 'overgeneralised'. The Himalaya is more than just the Middle Hills near Kathmandu and more than Uttarakhand. Yet these are the source locations for most data.

Since the original drafting of this work in 1989, a wave of confusion, caused by the 'uncertainties' of the IDE outsiders, first spread to affect large tracts of the academic literature and even some India's internal environmentalist analyses (Agarwal *et al.*, 1991). Here, after furious local debate highlighted the mismatch between the polemics of IDE complacency, and everyday experience of the Himalaya's problems (cf. Sagahal, 1991; Dogra, 1992), the idea quickly faded.

Outside India, the tide has also turned. A new synthesis is emerging. Its prescriptions are less extreme than the early environmentalist alarums, but less dismissive than those of the IDE. It is doubtful whether they are any more firmly founded in the realities of the problems in the Hills. However, today, foreign academics try to make their point by proving how reasonable and balanced are their new assessments. Once again, it is recognised that there is a problem in the physical environment of the Himalaya and that its impact may extend beyond the hills. It may or may not be universally disastrous in its impacts but it certainly has that capacity (Bruijnzeel, 1991). Meanwhile, retirement looms for the powerful leaders of the IDE's oft-called 'Mountain mafia'. Younger scientists with revisionist viewpoints wait in the wings.

Inside the Himalaya, the Chipko movement may have run out of steam. By 1990, in Kumaun, the energy of the early 1980s seemed to have dissipated. Chipko campaigners had moved on to new themes. Meanwhile, the situation in the forests and the local economy continued as before. As for the scientists of the Himalaya, for the majority, the bulk of their work seems destined to remain invisible to the outside world. However, patterns of scientific communication are changing. Nation-based publications are everywhere declining in significance. International journals feel more obliged to include contributions from workers outside the western mainstream. The World Wide Web promises freedom of information exchange to everyone who can 'log in'. It is always possible that the new world of electronic communication may shed some of the old prejudices.

However, there remains real social and political unrest in parts of the hills. As in eastern Europe's recent history, activism for political ends may, in part, have its roots in perceived environmental decline, itself triggered by outside interference. In the Himalaya, there remains the root problem within the local agroecosystem. In large tracts of the hills, agriculture is failing to provide subsistence to its people. There is an increasing pressure on the resource base of the traditional system and an increasing reliance upon the "money order" economy. The real problems of the Central Himalayan headwaters may be how to put the issue of agro-economic restructuring on a national agenda where it becomes a candidate for national resources, and it remains how to restrain a resource hungry nation from plundering increasingly accessible resources in its, no longer so isolated, headwaters.

Solutions to this problem have been conceived at all levels from the supranational to that of the individual villager. At present, a grass-roots approach which puts the farmer first seems to hold the best prospects for a long term solution (Shah, 1989, 1991). Environmental quality is ultimately secured by the efforts of those who live in that environment. However, such local activity can thrive only in an economic and institutional environment which is sympathetic. There are no simple answers, even to the question about whether international "aid" is beneficial or not. It could be argued that it is the outside world which gave celebrity status to the Himalayan environmentalists and their policies and thus enabled them to influence the policies of their national Governments. However, as the outside academic (IDE) community repeatedly pointed out, outsiders, like this author, really do not understand what is actually going on in the Himalaya.

Ultimately, environmental management policy in the Himalaya must depend upon the beliefs of the Himalayan people. Mahatma Gandhi stressed the importance of self-reliance and self-rule. At present, the definition of this term is a local political question which ranges Gandhi's own "Sarvodaya", village-based approach, against the top-down model preferred by Government. The front line of this debate is in the front ranges of the Himalaya. Here, deforestation, land degradation, and the decline of the local economy must be prevented. If not, the possibility remains that the environmental calamity so created will not remain nicely in the headwaters of the Himalayan ranges, but be carried down in floods, polluting sediments, and political unrest to the plains below.

REFERENCES

Agarwal, A., Chopra, R. and Sharma, K. 1982. *The State of India's Environment: A Citizen's Report.* New Delhi: Centre for Science and Environment: 192 pp.

Agarwal, A. and Narain, S. 1985. *The State of India's Environment: The Second Citizen's Report.* New Delhi: Centre for Science and Environment: 398 pp.

Agarwal, A., Chak, A. and Narain, S. 1991. *Floods, Flood Plains and Environmental Myths: The Third Citizens Report.* New Delhi: Centre for Science and Environment: 167 pp.

Agnihotri, Y., Dubey, L.N., and Dayal, S.K.N. 1985. Effect of vegetation cover on runoff from a watershed in Shivalik foothills. *Indian Journal of Soil Conservation* 13(1): 10–13.

Ahmed, S. 1985: The socio-political economy of deforestation in India. *University of East Anglia, Institute of Development Studies, Occasional Paper* 29: 1–84.

Ashish, M. 1979. Agricultural economy of the Kumaun hills: threat of ecological disaster. *Economic and Political Weekly* 14(25): 1058–1064.

Bahuguna, S.L. *et al.* 1981/3. Padyatra from Kashmir to Kohima. *Himalaya Man and Nature* 5(3): 3–6, 5(8): 3–9, 6(8): 6–7.

Bahuguna, S.L. 1985. Peoples response to ecological crises in the hill areas. pp. 217–226. In: Bandyopadhyay, J., Jayal, N.D., Schoetti, U., and Chhatrapati Singh (eds.): *India's Environment— Crises and Responses.* Dehra Dun, U.P.: Natraj: 314 pp.

Bandyopadhyay, J. and Shiva, V. 1987. Chipko: rekindling India's forest culture. *The Ecologist* 17(1): 26–34.

Bartarya, S.K. 1988. Geohydrological and geomorphological studies of the Gaula River Basin, District Nainital, with special reference to the problem of Erosion. Nainital, U.P. Kumaun University, Department of Geology, Unpublished Ph.D. thesis: 266 pp.

Bartarya, S.K. 1991. Watershed management strategies in Central Himalaya: the Gaula River Basin, Kumaun, India. *Land Use Policy* 8(3) 177–184.

Bartwal, P.S. 1987. Fuelwood consumption patterns in rural areas—a case study. *Journal of Tropical Forestry* 3(2): 136–141.

Bartwal, P.S. and Bhatt, A.B. 1987. Development and alternative sources of forest energy in Garhwal. In: Pangtey, Y.P.S. and Joshi, S.C. (eds), *Western Himalaya*. Nainital: Gyanodaya Prakashan: 747–754.

Bhatt, C.P. 1989. *Ecosystem of the Central Himalayas and Chipko Movement*. Gopeshwar, U.P.: Dashauli Gram Swarajya Sangh: 40 pp.

Bruijnzeel, L.A. 1990. *Hydrology of Moist Tropical Forests and Effects of Conversion: State of Knowledge Review*. Paris: UNESCO: IHP Humid Tropics Programme. 224 pp.

Bruijnzeel, L.A. 1991. Hydrological impacts of tropical forest conversion. *Nature and Resources (UNESCO)* 27(2) 36–46.

Bruijnzeel, L.A. and Brenner, C.N. 1989. *Highland-Lowland Interactions in the Ganges-Brahmaputra River Basin: A Review of the Published Literature*. Kathmandu, Nepal: ICIMOD Occasional Paper 11: 136 pp.

Chalise, R.S. 1986. Constraints on resources and development in the mountainous regions of South Asia. pp. 12–26. In: Joshi, S.C., Haigh, M.J., Pangtey, Y.P.S., Joshi, D.R. and Dani, D.D. (eds), 1986: *Nepal Himalaya: Geoecological Perspectives*. Nainital, India: H.R. Publishers: 525 pp.

Chaturvedi, O.P. 1985. Vegetation degradation in the U.P. hills. *Indian Farmers Digest* 18(4): 37–40.

Dav, H.U. and Kaul, V. 1986. Agroforestry: a possible way out of the increasing human predicament. *Journal of Tropical Forestry* 2(3): 177–182.

Dewan, M.L. 1990. Indian monsoons and Himalayan water conservation. *Himalaya: Man and Nature* 14(5): 17–21/14(6): 4–8.

Doctor, A.H. 1965. *Sarvodaya: A Political and Economic Study*. Delhi: Asia: 67–143.

Dogra, B. 1983. *Forests and People: A Report on the Himalaya*. New Delhi: Dogra: 8–20.

Dogra, B. 1992. *Forests, Dams and Survival in Tehri Garhwal*. New Delhi: Dogra: 85 pp.

Ehrich, C. 1989. The next flood knocking at the door: a foreigners view on the ecological crisis in India. Chapter 20. pp. 196–200. In: T.V. Singh and J. Kaur (eds) *Studies in Himalayan Ecology*. (Second Revised Edition: 1989) New Delhi: Himalayan Books: 286 pp.

Eckholm, E. 1976. *Losing Ground*. New York: 223 pp.

Fox, J. 1982. Managing public lands in Nepal. Madison, Wi., Development Studies: 11th Annual Conference on South Asia: unpublished typescript: 25 pp.

Glimour, D.A. 1988. Not seeing the trees for the forest: a reappraisal of the deforestation origin in two hill districts of Nepal. *Mountain Research and Development* 8(4): 343–350.

Government of India. 1980. *Report of the National Commission on Floods*. New Delhi: Department of Irrigation.

Gupta, P.N. 1979. *Afforestation, Integrated Watershed Management, Torrent Control and Landuse Development for U.P. Himalaya and Siwaliks*. Lucknow: Uttar Pradesh Forest Department: 67 pp.

Gupta, R.K., Kishore, V. and Ram Babu. 1982. Ecological background of Garhwalis—a study for environmental education and ecodevelopment. *Journal of Himalayan Studies and Regional Development* 5/6: 19–28.

Haigh, M.J. 1979a. Landslide sediment accumulations on the Mussoorie-Tehri road, Garhwal Lesser Himalaya. *Indian Journal of Soil Conservation* 7(1) 1–4.

Haigh, M.J. 1979b. Environmental geomorphology of the Landour-Mussoorie area (U.P.). *Himalayan Geology* 9(2): 657–668.

Haigh, M.J. 1981. Floods and erosion in North India. *Himalaya: Man and Nature* 5(1): 21–25, and 5(2): 4–9.

Haigh, M.J. 1982. A comparison of sediment accumulations beneath forested and deforested microcatchments, Garhwal Himalaya. *Himalayan Research and Development* 1(2): 118–120.

Haigh, M.J. 1984. Deforestation and disaster in northern India. *Land Use Policy* 1(3): 187–198. (Also reprinted as: Chapter 9. pp. 67–80. In: T.V. Singh and J. Kaur (eds.), *Studies in Himalayan Ecology.* (Second Revised Edition: 1989) New Delhi: Himalayan Books. 286 pp).

Haigh, M.J. 1988. Understanding Chipko: the Himalayan people's movement for forest conservation. *International Journal of Environmental Studies* 31 (2/3): 99–110.

Haigh, M.J. 1989. Water erosion and its control: case studies from South Asia. pp. 1–38. In: K. Ivanov and D. Pechinov (eds) *Water Erosion.* Paris: UNESCO Technical Documents in Hydrology IHP-III, Project 2.6: 140 pp.

Haigh, M.J. 1991. Reclaiming forest lands in the Himalaya: notes from Pakistan. In: Rajwar, G.S. (ed) *Advances in Himalayan Ecology.* New Delhi/Houston: Today and Tomorrow/Scholarly "Recent Researches in Ecology, Environment and Pollution". 6: 199–210.

Haigh, M.J., Rawat, J.S. and Bisht, H.S. 1990. Hydrological impact of deforestation in the Central
· Himalaya. *International Association of Hydrological Sciences, Publication* 190: (Hydrology of Mountainous Areas) 419–433.

Hamilton, L.S. 1985. Overcoming myths about soil and water impacts on tropical forest landuses. pp. 680–690. In: El-Swaify, S.A. *et al.* (eds.) *Soil Erosion and Conservation.* Ankeny, Iowa: Soil Conservation Society of America: 749 pp.

Hamilton. L.S. 1987. What are the impacts of Himalayan deforestation on the Ganges-Brahmaputra lowlands and delta? assumptions and facts. *Mountain Research and Development* 7(3): 256–263.

Hamilton, L.S. and Pearce, A.J. 1986. Biophysical aspects in watershed management. pp. 33–52. In: Easter, K.W., Dixon, J.A. and Hufschmidt, M.M. (eds), *Watershed Resource Management—An Integrated Framework with Case Studies from Asia and the Pacific.* Boulder, Co.: Westview Studies in Water Policy and Management 10: 236 pp.

Hedge, P. 1988. *Chipko and Appiko: How the People Save the Trees.* London: Quaker Peace and Service: 44 pp.

Himalaya Seva Sangh 1977–1995. *Himalaya, Man and Nature* (Journal). New Delhi: Rajghat.

Hopkins, N. 1983. Fodder situation in the hills of eastern Nepal, *APROSOC Occasional Paper* 2: 17 pp.

Ives, J.D. 1987. The Himalaya-Ganges problem: the theory of Himalayan environmental degradation: its validity and application challenged by recent research. *Mountain Research and Development* 7(3): 181–199.

Ives, J.D. 1988. Development in the face of uncertainty. pp. 55–75. In: Ives, J.D. (ed), *Deforestation: Social Dynamics in Watersheds and Mountain Ecosystems.* New York: Routledge/IUCN.

Ives, J.D. and Messerli, B. 1989. *The Himalayan Dilemma: Reconciling Development and Conservation.* New York: Routledge/UNU: 295 pp.

Ives, J.D., Hamilton, L.S. and O'Connor, K. 1989. Feasibility study for an International Mountain Research and Training Centre. *Mountain Research and Development* 9(2): 187–195.

Kattelmann, R. 1987. Uncertainty in assessing Himalayan water resources. *Mountain Research and Development* 7(3): 279–286.

Masrur, A. and Hanif, M. 1972. A study of surface runoff and sediment release in a Chir pine area. *Pakistan Journal of Forestry* 22(2): 113–142.

Mathur, H.N. and Sajwan, S.S. 1978. Vegetation characteristics and their effect on runoff and peak flow in small watersheds. *Indian Forester* 104(6): 398–406.

Metzner, R. 1993. The emerging ecological worldview. *Bucknell Review* 37(2): 163–172.

Misra, A. and Tripathi, S. 1978. *Chipko Movement: Uttarakhand Women's Bid to Save Forest Wealth.* New Delhi: Gandhi Peace Foundation: 40 pp.

Moddie, A.D. 1985. Development with desertification: U.P. Hills-Bhimtal's water system: facts, policies, lacuna, and responses, etc. *Central Himalayan Environment Association Bulletin* 2: 1–36.

Myers, N. 1984. The Himalayas: an influence on 500 million people. *Earthwatch* 3(7): 3–6.

Myers, N. 1986. Environmental repercussions of deforestation in the Himalayas. *Journal of World Forestry Resource Management* 2: 63–72.

Nautiyal, J.C. and Barbor, P.S. 1985. Forestry in the Himalayas—how to avert an environmental disaster. *Interdisciplinary Science Reviews* 10(1): 27–41.

Negi, S.S. 1983: Need for a conservation based rural energy strategy in the Himalayas. *Journal of Rural Development* 2(1): 134–143.

Nossin, J.J. 1971. Outline of the geomorphology of the Doon Valley, northern U.P., India. *Zeitschrift fur Geomorphologie* N.F. 12: 18–50.

Pandey, A.N., Pathak, P.C. and Singh, J.S. 1983. Water sediment and nutrient movement in forested and non-forested catchments in Kumaun Himalayas. *Forest Ecology and Management* 71: 19–29.

Pandey, U. and Singh, J.S. 1984. Energy flow relationships between agroforestry and forest ecosystems in central Himalaya. *Environmental Conservation* 11(1): 45–51.

Pant, D.C. 1983. Problems and prospects of tourism in the U.P. Himalayas. pp. 362–367, In: Singh, O.P. (ed) *The Himalaya: Nature, Man and Culture.* New Delhi: Rajesh.

Pant, H.B. and Rawat, G.S. 1989. Impact of industries and mining on water resources of the Doon valley. *Himalaya: Man and Nature* 12(8/9): 18–21.

Patnaik, S. and Ramakrishnan, P.S. 1989. Comparative study of energy flow through village ecosystems of two co-existing communities of Meghalaya in northeast India. *Agricultural Systems* 30: 245–267.

Poldane, M. 1987. Chipko movement: an overview. pp. 702–717. In: Pangtey, Y.P.S. and Joshi, S.C. (eds.), *Western Himalaya: Environment, Problems and Development.* Nainital: Gyanodaya Prakashan: 860 pp.

Prasad, R. 1983. Industrial potential of the U.P. Himalaya, pp. 198–204, In: Singh, O.P. (ed). *The Himalaya: Nature, Man and Culture.* New Delhi: Rajesh.

Ramsay, W.J.H. 1987. Deforestation and erosion in the Himalaya—is the link myth or reality. *International Association of Hydrological Sciences Publication* 167: 239–250.

Rai, S.C., Ahmad, A. and Rawat, J.S. 1989. *Environmental Degradation Due to Population Pressure, Migration and Settlement: A Case study of Central Himalaya.* Kosi, Almora: G.B. Pant Institute of Himalayan Environment and Development: 41 pp.

Rawat, A.S. 1983. *Garhwal Himalaya—A Historical Survey 1815–1947.* Delhi: Eastern Book Linkers: 105–136.

Rawat, M.S. and Rawat, J.S. 1994. Anthropogenic transformations of sheetwash erosion; experimental study in Kumaun Himalaya. (In preparation): 20 pp.

Rieger, H.C. 1976. Floods and droughts—the Himalaya and the Ganges Plain as an Ecological System. pp. 13–29. In: *Mountain Environment and Development.* Kathmandu: SATA.

Sastry, G. and Dhruva Narayanan, V.V. 1986. Hydrologic responses of small watersheds to different land uses in Doon Valley. *Indian Journal of Agricultural Sciences* 56(3): 194–197.

Sagahal, B. 1991. *'Himalayan Blunder' The third 'Citizens Report on the State of India's Environment':* A Review. Bombay: Sanctuary Magazine Publishers: 2 pp.

Shah, S.L. 1982. *Socioeconomic, Technological, Organisational and Institutional Constraints in the Afforestation of Civil, Soyam, Usar, and Waste lands for Resolving the Fuel Wood Crisis in the Hill Districts of Uttar Pradesh.* Almora: Vivekananda Laboratory for Hill Agriculture ICAR: 29 pp.

Shah, S.L. 1982. Ecological degradation and the future of agriculture in the Himalayas. *Indian Journal of Agricultural Economics* 37(1): 1–22.

Shah, S.L. 1986. Socioeconomic survey in village Chausali. *ICAR Vivekananda Laboratory for Hill Agriculture, Annual Report* 1986: 193–203.

Shah, S.L. 1991. Ecodevelopment in Uttarakhand: concepts and strategies, emerging lessons from action research in Khulgad micro watershed in Almora District: *Central Himalayan Environment Association Bulletin* 4: 1–28.

Sharma, K.C. 1981. Economic analysis of grassland development in Naurar watershed, Ramganga catchment in hills of Uttar Pradesh. Pantnagar: G.B. Pant University of Agriculture and Technology, unpublished M.Sc. thesis.

Shiva, V. 1986. Chipko: trees, water, and women. *Sanity (CND),* April: 30–33.

Shiva, V. 1987. Forestry myths and the World Bank. *The Ecologist* 17(4/5): 142–149.

Silas, R.A., Gaur, R.D. and Bartwal, P.S. 1987. Forest resources in the Raath region of Garhwal Himalaya. In: Pangtey, Y.P.S. and Joshi, S.C. (eds.), *Western Himalaya: Environment, Problems and Development.* Nainital: Gyanodaya Prakashan: 860 pp.

Singh, A.K. and Sharma, J.S. 1987. Women's contribution in hill agriculture. In: Pangtey, Y.P.S. and Joshi, S.C. (eds), *Western Himalaya: Environment, Problems and Development.* Nainital: Gyanodaya Prakashan: 860 pp.

Singh, J.S., Pandey, U. and Tiwari, A.K. 1984. Man and forest: a Central Himalayan case study. *Ambio* 13(2): 80–87.

Singh, T.V. and Kaur, J. 1986. The paradox of mountain tourism: case references from the Himalaya. *UNEP Industry and Environment* 9(1): 21–26.

Singh, V. and Naik, D.G. 1987. Fodder resources of central Himalaya. pp 223–241. In: Pangtey, Y.P.S. and Joshi, S.C. (eds) *Western Himalaya: Environment, Problems and Development.* Nainital: Gyanodaya Prakashan: 860 pp.

Subba Rao, B.K. Ramola, R.C. and Sharda, V.N. 1985. Hydrologic response of a mountain watershed to thinning: a case study. *Indian Forester* 111: 418–431.

Tiwari, A.K., Saxena, A.K. and Singh, J.S. 1986. Inventory of forest biomass for the Indian Central Himalaya. In: Singh, J.S. (ed), *Environmental Regeneration in the Himalaya.* Nainital: Gyanodaya Prakashan: 236–247.

Tiwari, A.K. and Singh, J.S. 1987. Analysis of forest landuse and vegetation in a part of the central Himalaya using aerial photographs. *Environmental Conservation 14(3):* 233–244.

Tiwari, S.C., Rawat, K.S. and Semwal, R.L. 1987. Extent and source of forest fire in the forest landscape of Garhwal Himalaya. pp. 553–564. In: Pangtey, Y.P.S. and Joshi, S.C. (eds.), *Western Himalaya: Environment, Problems and Development.* Nainital: Gyanodaya Prakashan: 860 pp.

Thompson, M. and Warburton, M. 1985. Uncertainty on a Himalayan scale. *Mountain Research and Development* 5(2): 115–135.

Upadhyay, H.C. 1987. Nature of rural indebtedness: a case study of scheduled castes in the Kumaun hills. pp. 799–803. In: Pangtey, Y.P.S. and Joshi, S.C. (eds.), *Western Himalaya: Environment, Problems and Development.* Nainital: Gyanodaya Prakashan: 860 pp.

Uttarakhand Seva Nidhi. 1986. *Workshop on 7th Five Year Plan for Kumaun Hills.* Nainital: Consul: 35 pp.

Valdiya, K.S. 1985a. Accelerated erosion and landslide prone zones in the Central Himalayan region. pp. 12–39. In: J.S. Singh (ed.), *Environmental Regeneration in the Himalaya.* Nainital: CHEA and Gyanodaya Prakashan.

Valdiya, K.S. 1985b. Himalayan tragedy—big dams, seismicity, erosion and drying up of springs in the Himalayan region. *Central Himalayan Environment Association Bulletin* 1: 1–24.

Valdiya, K.S. 1987. *Environmental Geology: The Indian Context. New* Delhi: Tata McGraw-Hill: 583 pp.

Valdiya, K.S. and Bartarya, S.K. 1989. Diminishing discharges of mountain springs in a part of Kumaun Himalaya. *Current Science* 58(8): 416–426.

Valdiya, K.S. and Bartarya, S.K. 1989. Problem of mass-movements in a part of Kumaun Himalaya. *Current Science* 58(9): 486–491.

Socioeconomic and Institutional Constraints on the Adoption of Soil Conservation Practices in the USA: Implications for Sustainable Agriculture

Ted L. Napier

ABSTRACT

Predicting the adoption of soil conservation practices by agriculturalists has proved problematic. Positive attitudes towards conservation do not, of themselves, always lead to changes in production practice. However, change does follow from the application of economic incentives and offers of technical assistance. Agriculturalists are strongly influenced by concerns about profitability and risk. Governments can affect profitability and the adoption of soil conservation practices either by targeting subsidies, as in the past, or by enforcing environmental standards. Concern about the development of this second option should encourage US agriculturalists to increase their own industry's efforts to reduce soil losses and also sediment and chemical pollution. The alternative to self-regulation may be regulation drawn up and enforced by non-farm interests.

Keywords: Soil conservation, socioeconomic constraints, sustainable agriculture, USA.

INTRODUCTION

Extensive research has been conducted during the past decade that has been focused on the adoption of soil and water conservation practices at the farm level, both domestically (Halcrow *et al.*, 1982; Lovejoy and Napier, 1986; Napier *et al.*, 1983) and internationally (Huszar and Cockrane, 1990; Napier, 1991; Napier, 1992; Napier *et al.*, 1994). Researchers have examined a multitude of attitudes and adoption behavior of land owner-operators under various environmental and socioeconomic conditions. Good statistical models have been formulated to predict attitudes of land owner-operators toward soil and water conservation. A number of significant variables have also been identified that affect awareness of environmental problems created by agricultural pollution. Unfortunately, prediction of adoption of conservation practices by agriculturalists has been shown to be problematic.

A number of factors have been shown to influence attitudes toward soil and water conservation at the farm level (Halcrow *et al.*, 1982; Lovejoy and Napier, 1986; Napier,

1990$_a$; Napier *et al.*, 1983, 1994). Some of the factors demonstrated to affect attitudes toward soil and water conservation issues and awareness of environmental problems are as follows: characteristics of the primary farm operator, characteristics of the farm enterprise, participation in farm programs, and access to information-education systems.

Research focused on the prediction of attitudes toward soil and water conservation reveals that environmental attitudes are influenced by well-designed intervention programs. Unfortunately, it has also been demonstrated that awareness of environmental problems and the development of positive attitudes toward environmental issues frequently *do not produce* changes in production practices of land owner-operators. Information-education programs are often implemented that have *little or no effect* on adoption of soil and water conservation practices. Well-designed and appropriately implemented information-education programs often result in the development of favorable attitudes among land owner-operators toward soil and water conservation issues, however, such efforts frequently have no effect on adoption behavior.

Many social, economic, and institutional factors have been examined in the context of adoption of soil and water conservation practices at the farm level (Camboni and Napier, 1993; Carlson and Dillman, 1988; Halcrow *et al.*, 1982; Lovejoy and Napier, 1986; Mueller *et al.*, 1985; Napier, 1990$_a$; Napier *et al.*, 1994; Napier and Brown, 1993; Napier and Camboni, 1993; Napier and Napier, 1991; Thomas *et al.*, 1990). Factors such as personal characteristics of land owner-operators, characteristics of the farm enterprise, psychosocial orientations of farmers, characteristics of soil and water conservation practices, community settings into which conservation programs are introduced, and many other variables have been evaluated as predictors of adoption behaviors at the farm level. While a number of variables have been shown to be significantly related with adoption behaviors and in expected directions, the amount of explained variance in practically all of the statistical models developed to date is very low. It is quite common for statistical models to explain no more than 5–15 percent of the variance in measures of adoption behavior. Explained variance of this magnitude indicates that the factors included as predictive factors in the statistical models are poor predictors of adoption behaviors. Findings from statistical models with low explained variance should not be used to develop conservation policy or to develop intervention strategies.

Variables that have been shown to consistently motivate land owner-operators to adopt soil and water conservation practices are government incentives and disincentives. Farmers respond to economic subsidies and to free technical assistance. Research has shown that the rapidity of adoption of soil and water conservation practices increases as the magnitude of economic and technical assistance subsidies increases.

During the conduct of scientific research focused on the adoption of conservation practices, a number of myths about why farmers accept or reject conservation technologies and techniques have been destroyed. Social scientists have discovered that many of the theoretical perspectives commonly employed to explain adoption of soil and water conservation practices at the farm level are relatively useless for prediction purposes. Repudiation of inappropriate explanations of adoption behavior is a very significant contribution of existing research.

Many of the findings derived from socioeconomic studies focused on adoption of conservation behaviors conducted in the United States (US) during the past decade have been difficult for conservation agencies to accept because the findings bring into serious question some of strategies presently being employed to facilitate adoption of conservation practices at the farm level. Existing research strongly suggests that some activities enacted by conservation agencies to motivate farmers to adopt do not have much effect in terms of modifying farm production systems. If the ultimate goal of soil and water conservation programs is to motivate land owner-operators to adopt and to continue using conservation practices on vulnerable land, then it will be necessary for change agencies to explore alternative intervention strategies.

The purpose of this paper is to discuss some of the social, economic, and policy factors that affect adoption of soil and water conservation practices at the farm level. The traditional diffusion paradigm (Rogers, 1983) is used to establish the rationale for selection of the variables discussed. The diffusion paradigm will also be used to demonstrate why a number of intervention strategies presently used by US conservation agencies came into being and why they continue to be used long after they should have been abandoned. Each factor selected for discussion will be examined in the context of why it *theoretically* should be important in explaining adoption of soil and water conservation behaviors of land owner-operators. Each factor will then be examined in the context of why it does not motivate land owner-operators to adopt soil and water conservation practices as expected. The paper is concluded with a brief discussion of future institutional actions that will probably be required to motivate land owner-operators to adopt soil and water conservation practices. The predictions are based on the assumption that the present sociopolitical environment in the US will result in significant reductions in the allocations of economic resources to the agricultural sector.

TRADITIONAL ADOPTION MODELS

Practically all contemporary strategies used to motivate land owner-operators to adopt soil and water conservation practices are some variation of the "Traditional Diffusion Model" (Rogers, 1993). The diffusion model is very well respected in a number of professional disciplines because it has been used successfully to predict adoption of technologies and techniques in many geographical regions of the world. While the traditional diffusion model has a number of serious limitations, it provides a useful starting point for analyzing adoption of technologies and techniques.

The traditional diffusion model basically asserts that human beings are reward seeking and punishment avoiding creatures that attempt to maximize rewards and to minimize costs in all social situations. The model argues that when people are exposed to information about innovations that are improvements over existing practices they will evaluate alternatives and form attitudes toward each option. Action options are assessed in the context of personal goals and situational conditions. Decisions are ultimately made about the merits of each alternative. The model posits that human beings will choose action options they believe will produce the most personal benefits at the least cost. Benefits are assessed using a variety of criteria such as savings in time, reduction in labor, savings in economics costs, returns to monetary investment, psychosocial satisfaction, and a number of other ways.

The traditional diffusion model asserts that characteristics of innovations are important considerations in the adoption decision-making process. An innovation that is perceived to be relevant to the needs of potential adopters, is simple to use, requires little investment in economic and human capital to effectively incorporate into existing farming systems, produces benefits in excess of costs, is certain to produce desired outcomes, is subject to direct observation in field demonstrations, is compatible with farming practices being used, and can be implemented in a sequential manner has a higher probability of being adopted than innovations with opposite characteristics.

The most important factor in the traditional diffusion model is access to information by potential adopters. The diffusion model asserts that potential adopters must be made *aware* of problems requiring corrective action and be *informed* of alternative actions required to resolve identified problems. In the context of soil and water conservation problems, land operators must be made aware of the environmental problems created by soil erosion and chemical pollution and have access to information about techniques and technologies to reduce or eliminate identified problems. It is assumed in the diffusion model that potential adopters are basically ignorant of existing problems and possible solutions. Provision of information from the perspective of the traditional diffusion model removes the most significant barrier to adoption of soil and water conservation practices at the farm level which is *operator ignorance.*

APPLICATION OF DIFFUSION CONCEPTS TO SOIL AND WATER CONSERVATION AT THE FARM LEVEL

Several factors highlighted in the brief overview of the traditional diffusion model were selected for discussion relative to why they often do not produce changes in soil and water conservation behaviors as expected. The factors selected for examination are as follows: awareness of soil and water conservation problems and solutions, attitudes toward soil and water conservation problems and solutions, characteristics of technologies and techniques to resolve soil and water resources problem, and public policies that affect adoption of conservation practices at the farm level. While public policies are not discussed in the traditional diffusion paradigm, the model recognizes that institutional factors can influence adoption decisions.

Awareness of soil and water conservation problems and alternative solutions have been posited by many proponents of the traditional diffusion model to be the most significant barrier to adoption of conservation practices at the farm level. While it is true that land operators cannot be expected to change production practices if they are not aware of environmental problems, there is considerable evidence that suggests land owner-operators in the US and many developed societies are quite knowledgeable of soil and water conservation problems on their land. Also, there is considerable evidence that US farmers and those in many developed societies are aware of technologies and techniques to resolve practically any erosion or water degradation problem that exists on their land. US land owner-operators are aware of conservation agencies and university faculties that can provide practically any type of information needed to resolve conservation problems.

Given this body of research findings, it should not be surprising that primary reliance on information strategies to motivate farmers to adopt soil and water conservation practices are seldom successful in bringing about *behavioral change.* Most land owner-operators in the US are already well informed about many environmental issues. It is highly likely that most land owner-operators in the US have access to more information than they can process and are provided more information than they require for decision making. It is also highly likely that farmers are much more aware of what is happening on their land than they are given credit for knowing. They may not know the exact amount of soil loss per acre, however, they are aware that erosion is occurring and are generally aware of the magnitude of the problem. Of greater importance in the decision-making process about adoption of conservation practices, many US farmers know the economic impacts of erosion on net farm income. These findings suggest that ignorance is not a characteristic of most successful production agriculturalists in the US and implies that reliance on information strategies will have only marginal effect on adoption behavior in the US.

Development of positive attitudes toward soil and water conservation practices is another factor that proponents of the traditional diffusion model argue affects adoption at the farm level. Diffusionists suggest that development of positive attitudes toward soil and water conservation and feelings of land stewardship among land owners will result in greater adoption of soil and water conservation practices. While it is true that negative attitudes toward soil and water conservation practices will impede adoption, evidence is overwhelming that most land owner-operators in the US place high value on conservation and are committed to stewardship of soil and water resources. This finding has been shown to be true for farmers who employ production practices that contribute to soil and water degradation. While education-information efforts may increase positive attitudes among potential adopters, such efforts often have only *marginal* impact on actual adoption behaviors. It is also possible to significantly increase positive attitudes toward conservation and stewardship via intervention programs and have *no effect* on conservation behaviors. *Characteristics of conservation technologies and techniques* can act as a major barrier to adoption of soil and water conservation practices at the farm level. Many conservation practices are technology-intensive, management-intensive, and chemical-intensive which effectively prevent a significant proportion of land owner-operators from adopting due to high costs. Many land owner-operators do not possess economic resources and management skills required to effectively implement and use conservation farming systems. It is often very costly in terms of time, money, and effort to acquire skills needed to effectively employ management-intensive conservation technologies and techniques.

Even simplistic conservation practices, such as grassed waterways, are often costly for land owner-operators to incorporate into existing production systems. Permanent removal of land resources from agricultural production constitutes a cost in terms of foregone return from set-side acreage. Operation of farm machinery around grassed waterways is another form of cost that farmers must be willing to accept or they will not implement such conservation programs.

Some conservation practices are economically very expensive to implement. For example, development of physical structures, such as terraces or rock dams, are usually very costly to construct and as a result many farmers do not adopt such erosion control systems. Unless conservation technologies and techniques are developed that are simple to use, inexpensive to implement, profitable in the short- and long-term, easily repaired and maintained, certain to produce expected outcomes, and will resolve environment problems, it is highly doubtful that a "technological solution" to soil and water conservation problems will ever be achieved.

A characteristic of conservation technologies and techniques that is extremely important in the decision-making process among US farmers is *profitability*. Land owner-operators are business persons who are motivated by profits (Miller, 1992). Profit motivated decision makers will not invest in any technology or technique unless it will produce profits for them in the short- and/or long-term. It is futile to ask farm operators to adopt conservation practices that will introduce higher levels of risk and uncertainty into their business operations unless they are adequately compensated for assuming additional risk. Unfortunately, most soil and water conservation practices are not profitable in the short-term and often not even in the long-term (Doster *et al.*, 1983; Mueller *et al.*, 1985). Most benefits from soil and water conservation investments accrue to nonfarm people in the form of improved water quality and reduced sediment load.

While most land owner-operators are not particularly concerned about off-site damages from agricultural sources, they also are not particularly concerned about on-site damages in the form of groundwater contamination by farm chemicals (Napier and Brown, 1993). Farmers are not eager to adopt production practices designed to protect groundwater resources from contamination even when they derive household water supplies from aquifers that will be contaminated by farm chemicals they apply (Napier and Brown, 1993).

Risk is another characteristic that potential adopters consider in the adoption of decision-making process. It is very difficult to motivate land owner-operators to adopt soil and water conservation practices, if adoption introduces higher levels of risk into the farm production system. A classic example is protection of groundwater resources from contamination by farm chemicals. It is well known that groundwater resources are subject to contamination by agricultural chemicals leached from chemical-rich soils by surface water percolating to subsurface aquifers. It is also recognized that one mechanism for protecting groundwater resources from chemical contamination is for land owner-operators to reduce fertilizer application rates. Many farmers could reduce fertilizer application rates considerably without adversely affecting production levels and in doing so protect groundwater quality. Unfortunately, many farmers perceive over-application of nutrients as constituting an "insurance" against adverse weather conditions or as a means of maximizing production when weather conditions are optimum. Small losses of nutrients in several growing seasons via over application are often more than adequately compensated in a single season when conditions are right for achieving maximum output. This is especially true when the unit cost of nutrients is low.

Many land owner-operators in the US are aware that the risk to health of family members due to contamination of drinking water from subsurface aquifers is extremely low (Baker *et al.*, 1994; Napier and Brown, 1993; Richards *et al.*, 1994). Therefore, farmers are less concerned about adverse health impacts than they are about loss of production and reduced net farm income. Land owner-operators are also aware that they can avoid adverse health impacts of contaminated groundwater by using commercial bottled water for household consumption. Farmers perceive that it is cheaper to purchase bottled drinking water than to reduce fertilizer application rates. This line of reasoning strongly suggests that it is highly unlikely that education programs designed to provide information to land owner-operators about potential impacts of fertilizer application rates on groundwater contamination will be effective in facilitating adoption of soil and water conservation at the farm level. This will probably be true even in geographic regions with extensive groundwater pollution problems.

Public conservation policies can affect adoption of soil and water conservation practices at the farm level (Napier, 1990$_b$) because government programs can reduce risk associated with adoption by providing subsidies and/or technical assistance to potential adopters. Implementation of costly conservation practices such as terraces, lagoons, rock dams, and other physical structures can be cost-shared by public agencies to facilitate adoption. Agency personnel can provide advice to potential adopters relative to the most cost effective means of resolving environmental problems.

Evidence of the effectiveness of public subsidies and the provision of technical assistance in achieving soil and water conservation goals is provided by the initial successes of national soil conservation programs implemented during the Dust Bowl era of US history. It is highly likely that significantly greater losses of productive farm land in the Midwest would have been destroyed without federal intervention during the 1930s and 1940s (Napier, 1990$_b$; Napier, 1988). Land owner-operators in the Midwest quickly implemented soil conservation programs that were recommended by federal conservation agents during the Dust Bowl era because they recognized that future production capacities of land resources were being destroyed by wind erosion. Many of the recommended practices were easy to implement and were inexpensive, however, they were very effective in reducing erosion of crop land. The on-site benefits were easily recognized by potential adopters and farmers quickly adopted recommended conservation technologies and techniques.

Early successes of the information-subsidy approach used during the Dust Bowl period established precedence for future conservation programs (Napier, 1990$_b$). Unfortunately, conditions changed over time that made the information-subsidy approach less useful and much less effective. The most significant factor that reduced the effectiveness of the information-subsidy strategy was the decline in return to investments in soil and water conservation efforts. Most of the inexpensive conservation practices have been implemented on much of the crop land in the US. Greater reduction in soil loss and improved water quality will be achieved at much higher cost. Many US farmers are very reluctant to internalize the costs of soil erosion because most of the damages from erosion are off-site costs. Land owner-operators are not motivated to adopt conservation practices by off-site damages because they are seldom directly affected by those types of costs.

Another issue that has changed in recent years is *access* to economic and technical assistance subsidies. Public resources in the US were allocated in the past to any land owner-operator who requested support. Recently, national policies were implemented that restricted allocation of designated conservation resources to land owner-operators with more serious environmental problems. Public conservation policies were targeted to specific types of land and water resource problems so that greater conservation benefits could be achieved. Evidence to date strongly suggests that targeting is probably one of the most successful soil and water conservation policy actions to be implemented in the past several decades.

Public conservation policies affect more than distribution of subsidies. Public conservation initiatives can affect *property rights.* Since property rights are allocated and protected by representatives of society, rights of ownership are arbitrary and can be changed. Significant modification of property rights can result from the development of implementation procedures designed to achieve national conservation objectives. The most recent example is the Conservation Title of the Food Security Act of 1985 (Napier, 1990$_a$). Individual property rights were changed significantly as public involvement in landuse determination was legitimized. For the first time in US History, land owner rights to use and/or abuse land and water resources were constrained by government action. The Conservation Title of the Food Security Act of 1985 ushered in a new era of public involvement in conservation and property rights. It is highly likely the Conservation Title and subsequent legislation will eventually produce soil and water conservation programs that will be successful in resolving environmental problems that have been basically immune to voluntary-type conservation efforts. It is also highly probable that future conservation programs will be much more regulatory in nature, since voluntary programs will not be able to motivate land owner-operators to internalize the costs of controlling erosion.

The Conservation Reserve Program (CRP) of the 1985 and 1990 farm bills is a good example of acceptance of conservation programs that contain elements of coercion. While land owners clearly prefer to participate in conservation programs that provide economic and technical assistance without constraints (Napier and Forster, 1982), the CRP demonstrates that land owner-operators will participate in conservation programs that have penalties attached. Approximately, 38 million acres of highly erodible farm land have been retired via the CRP through the 12 sign-up periods (Conrad, 1993). At least 22.6 million acres of the retired CRP land are "base acres" that would still be in commodity production and contributing to environmental degradation had the CRP not been implemented. While it is highly likely that the CRP program would have been much more successful in reducing environmental degradation had it been implemented using different criteria and objectives (Reichelderfer and Boggess, 1988), the program has been successful using the number of acres of highly erodible land as an indicator of success.

While the CRP has been well received by land owner-operators in the US, it must be recognized that the success of the program has been achieved at a very substantial economic cost. The direct economic cost of the CRP in terms of yearly rents is now about 2 billion dollars per year. The administrative costs of the program are extremely high and are not included in the 2 billion dollar figure.

FUTURE DIRECTIONS OF SOIL AND WATER CONSERVATION POLICY INITIATIVES

It is highly likely that present conservation policy initiatives will be continued into the future *assuming* no constraints on resources to maintain established programs. However, it is highly doubtful that US society will continue to allocate existing levels of economic and human resource support to address soil and water conservation programs. Government spending is *almost certain* to be reduced in the near future which means many of the existing soil and water conservation initiatives will not be renewed. Public resources presently allocated to land owner-operators to control soil erosion and to improve water quality will probably be reallocated to other problems such as upgrading urban waste water treatment facilities, reclamation of contaminated industrial sites, investment in air pollution abatement facilities, investment in alternative fuels and energy research and development, and other such programs. Claimants on limited public resources will be numerous and politically more powerful than the agricultural sector.

Unfortunately, it is also highly likely that society will expect the agricultural sector to maintain present levels of resource protection even with significant reductions in funding levels. It is also quite possible that society will expect higher levels of environmental protection from agriculturalists. Should either or both of these predictions prove to be correct, the implications for land owner-operators are extremely significant. Farmers will be under severe pressure from representatives of society to internalize most of the costs of soil erosion control and water quality improvement programs. This means that subsidies to encourage adoption of soil and water conservation practices at the farm level will probably be reduced or eliminated.

If economic incentives are reduced as means of motivating land owner-operators to adopt conservation practices and if public opinion demands maintenance of high standards of environmental quality, it is highly likely that land owner-operators will be required to either internalize the costs of improving environmental quality or they will be subject to coercive forces in the form of regulations. Based on social science findings (Napier and Forster, 1982), it is highly probable that land owner-operators will not adopt soil and water conservation practices voluntarily, especially conservation practices that require investment of limited human and economic resources. While agriculturalists will lobby for exemption from environmental standards as they have been many times in the past, it is doubtful that any segment of society will be exempt from environmental standards in the future. This suggests that agriculture will be subject to much closer scrutiny than in the past.

If the agricultural sector does not assume a greater proportion of the costs of controlling agricultural pollution in the future, it is almost certain the industry will be regulated. Regulation can be achieved in a variety of ways such as the following: mandated nutrient application rates, quotas on commodities to prevent maximization of output, mandated use of specific production systems, periodic random inspection of production systems to estimate pollution potential, permanent retirement of highly erodible land via purchase of production rights, permanent retirement of highly erodible land from food and fiber production without compensation, establishment of environmental quality standards with land owner-operators provided the freedom to use any production system they desire as long as the system meets standards (burden of proof should be on the operator and not the state), and many other options.

Precedence has already been established via the Conservation Title of the 1985 and 1990 Farm Bills. The US Government has demonstrated its willingness to change property rights to achieve national conservation objectives. If society elects to implement cost-savings programs by reducing subsidies to the agricultural sector, it is almost certain to ultimately lead to wholesale use of disincentive systems. It is highly probable that the "stick" will replace the "carrot and stick" approach that is presently employed to motivate farmers to adopt conservation practices. In a coercive conservation policy environment most social, economic, and educational-informational approaches of the past will be of little utility as motivators of adoption behaviors.

Evidence to date suggests that the US Society can probably achieve national environmental quality goals using a coercive approach, however, the social cost of achieving conservation goals using such a strategy will undoubtedly be severe. The time remaining for the agricultural sector to act to prevent regulation of the industry is very limited. If land owner-operators in the US wish to retain control of their industry, they must act within a very short time frame to control soil erosion and reduce use of chemical inputs. If they do not act quickly, their future will probably be determined for them. Most of the principal actors in the development and implementation of a regulatory approach will be nonfarm interests and they will have little sympathy with the problems of farmers. The outcome of such a situation will undoubtedly be negative from the perspective of land owner-operators.

REFERENCES

Baker, David B., Wallrabenstein, Laura, K. and Richards, R. Peter. 1994. Well Vulnerability and Agrochemical Contamination: Assessments from a Voluntary Well Testing Program. *Proceedings of the Fourth National Conference on Pesticides: New Directions in Pesticide Research, Development and Policy.* Diana Weigmann (ed). Blacksburg, Virginia: Water Resources Center of VPISU.

Batie, Sandra. 1986. Why Soil Erosion: A Social Science Perspective. In: *Conserving Soil: Insights from Socioeconomic Research,* S.B. Lovejoy and T.L. Napier (eds.). Ankeny, Iowa: Soil and Water Conservation Society Press. pp. 3–14.

Camboni, Silvana M. and Napier, Ted L. 1993. Factors Affecting Use of Conservation Farming Practices in East Central Ohio. *Agriculture, Ecosystems and Environment* 45(1): 79–94.

Conrad, Daniel. 1993. Personal conversation with the assistant state conservationist on March 10, 1993.

Carlson, John E. and Dillman, Don A.1988. The Influence of Farmers' Mechnical Skill on the Development and Adoption of a New Agricultural Practice. *Rural Sociology* 53(2): 235–245.

Doster, D.H., Griffith, D.R., Manning, J.V. and Parsons, S.D. 1983. Economic Returns from Alternative Corn and Soybean Tillage Systems in Indiana. *Journal of Soil and Water Conservation* 38(6): 504–508.

Ervin, David. 1986. Constraints to Practicing Soil Conservation: Land Tenure Relationships. In: *Conserving Soil: Insights from Socioeconomic Research,* Stephen B. Lovejoy and Ted L. Napier (eds.). Ankeny, Iowa: Soil and Water Conservation Society Press. pp. 36–62.

Halcrow, Harold G., Heady, Earl O. and Cotner, Melvin L. (eds.). 1982. *Soil Conservation Policies, Institutions, and Incentives.* Ankeny, Iowa: Soil and Water Conservation Society Press.

Huszar, Paul, C. and Cockrane, H.C. 1990. Constraints to Conservation Farming in Java's Uplands. *Journal of Soil and Water Conservation* 45(3): 420–423.

Lovejoy, Stephen B. and Napier, Ted L. (eds.). 1986. *Conserving Soil: Insights from Socioeconomic Research.* Ankeny, Iowa: Soil and Water Conservation Society Press.

Miller, William L. 1982. The Farm Business Perspective and Soil Conservation. In: *Soil Conservation Policies, Institutions, and Incentives,* H.G. Halcrow, E.O. Heady, and M.L. Cotner (eds.) Ankeny, Iowa: Soil and Water Conservation Society. pp. 151–162.

Mueller, D.H., Klemme, R.M. and Daniel, T.C. 1985. Short- and Long-term Cost Comparisons of Conventional and Conservation Tillage Systems in Corn Production. *Journal of Soil and Water Conservation* 40(5): 466–470.

Napier, Ted L. 1988. Anticipated Changes in Rural Communities due to Financial Stress in Agriculture: Implications for Conservation Programs. In: *Impacts of the Conservation Reserve Program in the Great Plains,* J.E. Mitchell (ed.). Fort Collins, Colorado: U.S. Forest Service and the Range Management Society of America. pp. 84–90.

Napier, Ted L. (ed.). 1990$_a$. *Implementing the Conservation Title of the Food Security Act of 1985.* Ankeny, Iowa: Soil and Water Cconservation Society Press.

Napier, Ted L. 1990$_b$. The Evolution of U.S. Soil Conservation Policy: From Voluntary Adoption to Coercion. In: *Soil Erosion on Agricultural Land,* J. Boardman, I. Foster, and J. Dearing (eds.). London: Wiley Publishers. pp. 627–644.

Napier, Ted L. 1991. Factors Affecting Acceptance and Continued Use of Soil Conservation Practices in Developing Societies: A Diffusion Perspective. *Agriculture, Ecosystems and Environment* 36: 127–140.

Napier, Ted L. 1992. Property Rights and Adoption of Soil and Water Conservation Practices. In: *Conservation Policies for Sustainable Hillslope Farming,* S. Arsyad, I. Amien, T. Sheng, and W. Moldenhauer (eds.). Ankeny, Iowa: Soil and Water Conservation Society Press.

Napier, Ted L. and Camboni, Silvana M. 1988. Attitudes Toward a Proposed Government Sponsored Soil Conservation Program. *Journal of Soil and Water Conservation* 43(2): 186–191.

Napier, Ted L. and Napier, Anthony S. 1991. Perceptions of Conservation Compliance Among Farmers in a Highly Erodible Area of Ohio. *Journal of Soil and Water Conservation* 46(3): 220–224.

Napier, Ted L., Scott, D.F., Easter, K.W. and Supalla, R. (eds.). 1983. *Water Resources Research: Problems and Potentials for Agriculture and Rural Communities.* Ankeny, Iowa: Soil and Water Conservation Society.

Napier, Ted L. and Forster, D. Lynn. 1982. Farmer Attitudes and Behavior Associated with Soil Erosion Control. In: *Soil Conservation Policies, Institutions, and Incentives*, H.G. Halcrow, E.O. Heady, and M.L. Cotner (eds.). Ankeny, Iowa: Soil and Water Conservation Society. pp. 137–150.

Napier, Ted L. and Brown, Deborah E. 1993. Factors Affecting Attitudes Toward Ground Water Pollution Among Ohio Farmers. *Journal of Soil and Water Conservation* 48(5): 432–438.

Napier, Ted L. and Camboni, Silvana M. 1993. Use of Conventional and Conservation Practices Among Farmers in the Scioto River Basin of Ohio. *Journal of Soil and Water Conservation* 48(3): 231–237.

Napier, Ted L., Camboni, Silvana M. and El-Swaify, Samir A. (eds.) 1994. *The Socicoeconomics of Soil and Water Conservation: An International Perspective*. Ankeny, Iowa: Soil and Water Conservation Society Press.

Reichelderfer, K. and Boggess, William G. 1988. Government Decision Making and Program Performance: The Case of the Conservation Reserve Program. *American Journal of Agricultural Economics* 70: 1–11.

Richards, R. Peter, Baker, David B., Christensen, Brian and Tierney, Dennis. 1994. Atrazine Exposures Through Drinking Water: Exposure Assessments for Ohio, Illinois, and Iowa. In: *Proceedings of the Fourth National Conference on Pesticides: New Directions in Pesticide Research, Development and Policy*, Diana Weigmann (ed.). Blacksburg, Virginia Water Resources Center.

Roger, E.M. 1983. *Diffusion of Innovations*. New York: The Free Press.

Swanson, Louis E., Camboni, Silvana M. and Napier, Ted L. 1986. Barriers to the Adoption of Soil Conservation Practices on Farms. In: *Conserving Soil: Insights from Socioeconomic Research*, Stephen B. Lovejoy and Ted L. Napier (eds.). Ankeny, Iowa: Soil and Water Conservation Society Press. pp. 108–120.

Thomas, John K., Ladewig, Howard and McIntosh, Wm. Alex. 1990. The Adoption of Integrated Pest Management Practices Among Texas Cotton Growers. *Rural Sociology* 55(3): 395–410.

Printed in India